The Symbiotic Relationship Between FHIR and AI

Managers' Guide to Understanding FHIR and AI Integration

Claude Louis-Charles, Phd

i

Cybersoft Publishing LLC

Fort Washington, MD 20744

Drclaude.net

First Edition March 2026

Table of Contents

Introduction

Healthcare organizations didn't choose to become AI-driven—they simply woke up one day to find themselves surrounded by algorithms, vendor tools, and automated decisions they were suddenly accountable for. As your manuscript states, *"Most organizations didn't decide to become 'AI organizations.' It crept up on them."* That creeping reality is now a strategic turning point: leaders must understand how AI works, how it fails, and how to keep it safe. This book opens by confronting that pressure head-on, giving managers the clarity they've been missing.

The introduction makes one truth unmistakable: AI cannot be trusted unquestioningly, nor can it be governed casually. High-stakes environments—clinical care, claims adjudication, care coordination, and population health—require **Human-in-the-Loop (HITL)** oversight. As you write, HITL ensures that humans *"actively review, validate, override, and improve"* AI decisions. This book shows managers how to build that oversight without slowing down innovation. Readers will immediately understand that AI safety is not a technical detail—it is a leadership responsibility.

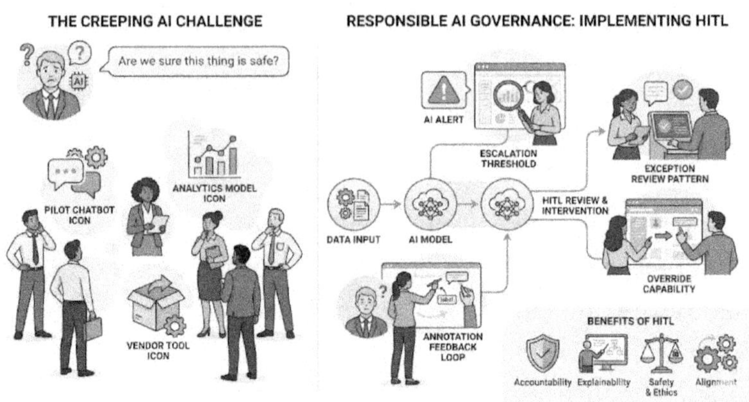

But the introduction also reveals the deeper challenge: AI is only as good as the data beneath it. Fragmented EHRs, inconsistent coding, and legacy standards make modern intelligence nearly impossible to scale. That is why this book positions **FHIR** as the missing link. FHIR is not just another standard—it is the foundation that turns scattered clinical data into structured, machine-readable intelligence. When paired with AI, it unlocks what you call *"intelligent interoperability, where data flows seamlessly, and systems learn continuously.* This framing gives readers a powerful reason to continue: the future of healthcare belongs to leaders who understand this symbiosis.

By the end of the introduction, the reader knows exactly why they must keep going. This book will teach them how to evaluate their data environment, build trustworthy AI workflows, design HITL guardrails, and lead their organization through the transformation already underway. It promises not hype, but operational clarity—something every healthcare manager urgently needs. The next chapters deliver the roadmap.

Claude Louis-Charles, PhD

1 Why FHIR + AI Is the Future of Healthcare

Every healthcare organization today operates under the same paradox: it generates enormous volumes of clinical data, yet struggles to use that data to make better decisions in the time it takes for those decisions to matter. A physician enters an order, a lab returns a result, and a payer adjudicates a claim. Each event produces structured or semi-structured data that, in theory, could inform the next clinical action. In practice, that data sits in silos, arrives too late, or arrives in a format that downstream systems cannot interpret without custom translation.

At the same time, artificial intelligence has matured from an academic curiosity into a practical toolkit. Machine learning models can detect early signs of sepsis, flag drug-drug interactions, predict hospital readmissions, and identify patients who are likely to benefit from a care management program. The tools exist. The computing power exists. The investment appetite exists. What has been missing — consistently, across health systems, payers, and vendors — is the data infrastructure required to make AI work reliably and safely at scale.

HL7 FHIR — Fast Healthcare Interoperability Resources — is not the only answer to that infrastructure problem, but it is the most important standard available today for solving it. FHIR defines a common language for healthcare data: a set of resource types, API conventions, and terminology bindings that allow disparate systems to exchange information in a way that machines can actually read and act on. When AI and FHIR operate together, something more powerful than either technology alone becomes possible: intelligent interoperability, where data flows seamlessly, and systems learn continuously from that flow.

This chapter establishes the foundational case. It explains why the healthcare data problem is worse than most organizations acknowledge, why AI needs structured data to deliver on its promise, and why FHIR is the missing link that turns AI pilots into production systems. Most importantly, it frames the relationship between FHIR and AI not as a one-way dependency but as a genuine symbiosis — each standard making the other more valuable. For any manager responsible for technology strategy, clinical operations, or enterprise data, this framing should shape every investment and build decision that follows.

1.1 The Healthcare Data Problem

Healthcare data is not scarce. The average hospital encounter generates hundreds of discrete data points — vital signs, lab values, medication administrations, nursing assessments, imaging studies, billing codes, and patient-reported information. A mid-sized health system with 500 beds may process tens of millions of such records per year. The U.S. healthcare system as a whole generates petabytes of clinical data annually. The problem is not volume. The problem is that this data is fragmented, inconsistent, inaccessible in real time, and formatted in ways that resist automated processing.

1.1.1 Fragmentation Across EHRs, Payers, Labs, and Devices

A patient with a chronic condition does not receive care in a single place. They see a primary care physician, visit a specialist, get labs drawn at an independent laboratory, fill prescriptions at a pharmacy, wear a continuous glucose monitor, and periodically interact with their health plan for prior authorizations and care management. Each of these touchpoints lives in a different system: a major EHR platform, a payer claims database, a laboratory information system, a pharmacy management system, a device manufacturer's cloud, and a population health management tool. None of these systems was designed to talk to the others natively.

The result is a clinical record that is rarely complete in any single place. When a patient arrives in the emergency department, the treating physician typically has access only to what that health system has on file. If the patient was seen last week at an affiliated urgent care clinic on a different EHR instance, that record may not be accessible. If the patient was recently discharged from a competing hospital, that information is almost certainly not available without a manual records request. The physician makes clinical decisions with an incomplete picture — a situation that is not the exception but the rule.

For AI systems, this fragmentation is catastrophic. A readmission prediction model trained on data from one health system will have systematically incomplete feature sets for patients who receive substantial care elsewhere. A medication reconciliation algorithm cannot reconcile what it cannot see. A sepsis alert system trained on a particular EHR's data structure will fail silently when deployed against a different EHR that encodes the same clinical concepts differently. Fragmentation does not just inconvenience humans — it actively corrupts the training data and inference inputs that AI depends on.

Diagram 1.1 – The Fragmented Healthcare Data Landscape

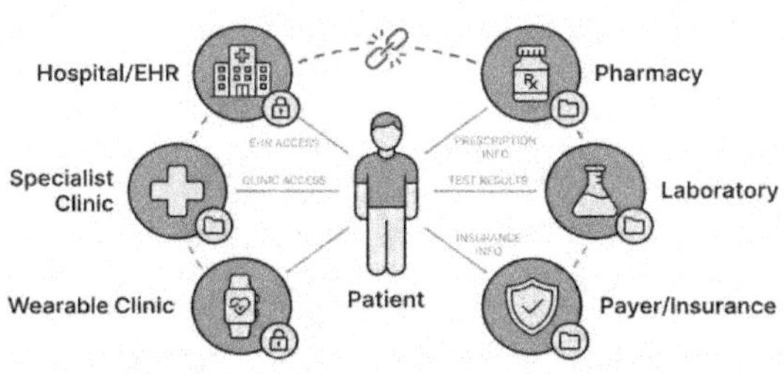

The fragmentation problem is compounded by the fact that even within a single health system, data is rarely unified. A large academic medical center may run multiple EHR instances across its enterprise — different versions, different configurations, different custom fields — stitched together by interface engines built over the years in response to acquisitions, mergers, and departmental preferences. The technical debt accumulated in these interface layers is often invisible to clinical and administrative leaders until a major upgrade, a migration, or an AI initiative forces it into view.

From a management perspective, fragmentation represents both a clinical risk and a strategic liability. Every AI initiative that requires cross-system data integration will encounter this fragmentation as a primary obstacle. Understanding the scope of that fragmentation before committing to an AI roadmap is not optional — it is foundational due diligence. A realistic data inventory, mapping which systems hold which clinical concepts and how they communicate, is among the highest-value activities a health IT leadership team can undertake.

1.1.2 Administrative Burden and Clinician Burnout

Fragmented data does not stay in databases. It flows upstream into the work lives of clinicians and administrative staff. Physicians spend significant portions of their working hours on documentation, prior authorization requests, referral coordination, and inbox management — tasks largely due to systems that cannot share information effectively. When a referral requires a physician's office to fax records to a specialist because there is no interoperable connection, someone must pull those records, organize them, and send them manually. When a prior authorization requires detailed clinical documentation that the payer's portal cannot accept electronically, a nurse or medical assistant transcribes it by hand.

The American Medical Association and other professional bodies have consistently documented that administrative burden is one of the top drivers of physician burnout. The downstream effects are well-established:

burnout correlates with higher rates of medical errors, higher turnover, reduced patient satisfaction, and lower quality of care. For health system executives, this is not a wellness issue — it is a patient safety issue and a financial issue. When experienced clinicians leave the profession or reduce their hours, the cost of replacement and the risk of care gaps are substantial.

AI has the potential to dramatically reduce administrative burden. Natural language processing can draft clinical notes, extract data from unstructured text, and pre-populate prior authorization forms. Intelligent automation can route referrals, flag incomplete orders, and surface relevant prior records at the point of care. But none of these capabilities can be deployed reliably unless the underlying data infrastructure supports them. An AI system that automates prior authorization forms still needs to retrieve the correct clinical data from the appropriate source in real time. Without interoperability, automation moves the problem rather than eliminating it.

1.1.3 Why Legacy Standards Cannot Support Modern Intelligence

Healthcare interoperability is not a new problem. The industry has been attempting to solve it for decades through a succession of standards and technologies. HL7 v2. x, the messaging standard introduced in the late 1980s, became the backbone of clinical data exchange and remains in use today. HL7 v3 and its Clinical Document Architecture (CDA) sought to provide greater semantic precision in clinical documents. The Consolidated CDA (C-CDA) became the basis for Meaningful Use attestation. Each generation of standards represented genuine progress, but each also carried forward limitations that compound over time.

HL7 v2. X was designed for point-to-point messaging between systems within a single organization. Its segment-based format is verbose, inconsistently implemented across vendors, and difficult to query. A single HL7 v2 ADT message for a patient admission may contain dozens of segments, only a fraction of which are populated in any given implementation. Vendors have extended the standard with proprietary fields, creating dialects that are technically compliant but practically incompatible. When two health systems attempt to exchange HL7 v2 messages, they almost always require custom mapping work on both ends.

CDA and C-CDA improved semantic richness by introducing XML-based document structures with standardized sections and coded entries. But they were designed as document exchange formats, not as real-time data APIs. Retrieving a specific piece of information from a C-CDA document — say, the patient's most recent HbA1c value — requires parsing an entire XML document and navigating a complex template hierarchy. For AI systems that need to query specific clinical facts across large patient populations in real time, this approach is not viable. Legacy standards were built for a document-centric world. AI requires a data-centric world.

The workflow-level impact of this mismatch is felt every time a health system attempts to build an AI feature on top of legacy infrastructure. Engineering teams spend the majority of their project timelines on data extraction, normalization, and validation — not on the AI logic itself. The ratio of data preparation to model development work in healthcare AI projects consistently runs at 3:1 or higher. That ratio is a direct consequence of legacy standards that were never designed to support machine learning pipelines.

1.1.4 The Cost of Inconsistent, Inaccessible Data

The financial cost of poor healthcare data quality is difficult to calculate precisely, but credible estimates are staggering. Research published by the Journal of AHIMA and other health information management bodies suggests that poor data quality contributes to billions of dollars annually in claim denials, duplicate testing, medication errors, and inefficient care coordination. These costs are not abstract — they appear in specific line items on health system income statements: denied claims that require rework, duplicate imaging orders that could have been avoided with access to records, and readmissions driven by inadequate discharge planning.

For AI initiatives specifically, poor data quality has a compounding cost structure. A model trained on inconsistent data will produce inconsistent predictions. A model deployed against incomplete data will generate alerts and recommendations that clinicians learn to distrust. Once a clinician's trust in an AI system erodes, it is extraordinarily difficult to rebuild — even if the underlying model is subsequently improved. The phrase 'garbage in, garbage out' is a cliché because it is empirically accurate, and in healthcare, the stakes of that garbage output are clinical, legal, and reputational.

For managers evaluating AI investments, this creates a practical decision framework: before investing in AI capabilities, assess the data quality and accessibility of the underlying clinical data. An honest assessment will reveal specific gaps — missing values, inconsistent coding, incomplete patient matching across systems — that must be addressed before any AI initiative can achieve operational clarity in production. Organizations that skip this assessment consistently underestimate project timelines and overestimate early model performance.

Diagram 1.2 – The Cost Cascade of Poor Data Quality

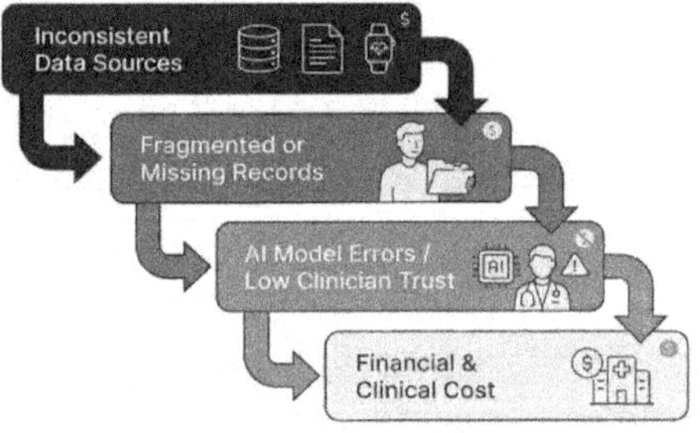

Inaccessibility is a distinct but related problem. Data that exists but cannot be retrieved in time to influence a clinical decision has no operational value. A patient's complete allergy list, stored in a system that cannot respond to a real-time API query, is effectively invisible at the point of care. FHIR addresses this inaccessibility directly through its RESTful API design — but realizing that benefit requires health systems to have deployed FHIR endpoints that are actually populated with accurate, current data. The standard creates the channel; data governance and infrastructure investment create the content.

1.2 The Rise of AI in Healthcare

Artificial intelligence in healthcare is not new, but its practical utility has changed dramatically over the past decade. What began as expert systems with hand-coded clinical rules has evolved into a diverse ecosystem of machine learning models, deep learning architectures, and large language models capable of processing clinical text, medical imaging, genomic sequences, and real-time sensor data. Understanding that evolution — and understanding what drives AI success or failure — is essential context for any manager making decisions about AI adoption.

1.2.1 From Rule-Based Systems to Machine Learning

The first generation of clinical decision support systems operated on explicit rules written by clinical experts: if potassium is below 3.0 and the patient is on digoxin, fire an alert. These systems were interpretable, auditable, and clinically grounded. They were also brittle. Every clinical scenario required a rule. Edge cases fell through the cracks. Alert fatigue emerged as systems generated hundreds of notifications per clinician per shift, the vast majority of which were overridden or ignored. Rule-based systems traded precision for recall — they fired alerts broadly to avoid missing serious cases. Still, in doing so, they degraded the signal-to-noise ratio to the point where alerts became background noise.

Machine learning introduced a fundamentally different paradigm: instead of explicitly encoding rules, train models on historical data and let the patterns in that data drive predictions. Early applications focused on structured clinical data — lab values, vital signs, medication histories, diagnoses — because that data was already stored in EHRs in queryable formats. Models predicting sepsis, readmission, and length of stay emerged from academic medical centers and demonstrated meaningful improvements in predictive accuracy compared to rule-based alternatives.

Deep learning extended these capabilities to unstructured data, including radiology images, pathology slides, clinical notes, echocardiograms, and retinal photographs. A convolutional neural network trained on chest X-rays can detect pneumonia with radiologist-level accuracy. A model trained on fundus photographs can identify diabetic retinopathy at scale in community settings where ophthalmologists are not available. Large language models, trained on vast corpora of clinical text, can extract structured clinical facts from unstructured notes, generate discharge summaries, and answer clinical questions in plain language.

The trajectory is clear, and the capabilities are real. But capability in a research setting and reliable deployment in a production clinical environment are not the same thing. The gap between them is almost always a data problem — and specifically, a data infrastructure problem.

1.2.2 Why AI Needs Clean, Structured, Interoperable Data

Machine learning models are statistical functions that map inputs to outputs based on patterns learned from training data. The quality of those inputs bounds the quality of those outputs. This is not a controversial claim — it is a mathematical constraint. A model trained on incomplete patient records will learn to make predictions in the presence of systematic incompleteness. If it is subsequently deployed in an environment where records are more complete—or less complete —its predictions will be miscalibrated. This is the training-deployment distribution shift problem, and it is one of the most common reasons AI models that perform well in research fail in production.

Structured data matters because machine learning models, at their core, require numerical inputs. Clinical text, imaging files, and waveform data can be converted to numerical representations through various encoding techniques. Still, structured data — discrete values with defined types, units, and coding systems — is directly usable. When a lab result is stored as a FHIR Observation resource with a LOINC code, a numerical value, and a unit, it is immediately machine-readable without preprocessing. When the same result is embedded in a narrative note or encoded in a proprietary EHR field, it must be extracted, normalized, and validated before it can feed a model.

Interoperability matters because the clinical picture of any given patient is distributed across systems. An AI model predicting cardiovascular risk needs medication data, lab data, vital signs, diagnoses, and social determinants of health. Each of these data types may live in a different system. A model that can only access the data available in one system will produce predictions based on an incomplete feature set — and the incompleteness will be systematically biased toward patients who happen to receive more of their care within that system. Interoperability is not a technical nicety; it is an equity and accuracy requirement for clinical AI.

Diagram 1.3 – What AI Needs: The Data Quality Stack

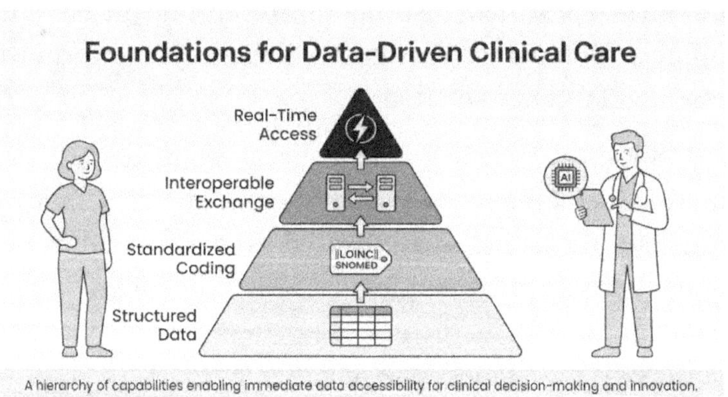

A hierarchy of capabilities enabling immediate data accessibility for clinical decision-making and innovation.

Real-time access is a third dimension of the data quality requirement that is often underweighted. Many AI use cases in healthcare are time-sensitive: sepsis prediction is most valuable if it fires 12 hours before clinical deterioration, not 12 hours after. A medication safety alert is useful when a physician is writing an order, not when it arrives after the order has been signed. These use cases require AI systems to query current clinical data rather than historical snapshots. That requires live, queryable data interfaces — precisely what FHIR's RESTful API architecture was designed to provide.

For managers, the practical implication is this: before evaluating which AI models to purchase or build, evaluate whether your data infrastructure can support real-time, structured, interoperable data access. An AI vendor's performance claims are almost always based on training and validation datasets that were curated, cleaned, and often sourced from a small number of high-performing academic medical centers. The question is not whether the model performed well in that environment — the question is whether your data environment is comparable.

1.2.3 Examples of AI Failing Due to Poor Data Quality

The research literature and post-market surveillance reports document numerous cases where AI systems failed in production due to data quality issues. These cases are instructive not as cautionary tales about AI in general, but as specific illustrations of what happens when capable models meet inadequate data infrastructure.

One widely studied case involved a sepsis prediction model deployed across multiple hospitals in a large health system. The model had demonstrated strong predictive performance in retrospective validation. In prospective deployment, however, performance degraded significantly at sites using a different version of the EHR, where laboratory data were encoded in a different field and vital signs were charted on a different schedule. The model had learned patterns specific to the data-encoding conventions of one

EHR configuration, not to the underlying clinical physiology. The data structure was not standardized across sites, so the model's inputs effectively served as different variables.

Another illustrative case involved a clinical NLP system designed to extract social determinants of health from clinical notes and populate structured fields for population health analytics. The system performed well on notes generated by physicians who dictated using standard templates. It performed poorly on notes generated by nurses, therapists, and case managers, whose documentation styles varied substantially. Because social determinants data were disproportionately documented by non-physician clinicians, the system systematically underperformed for the patient population in which social determinants were most clinically relevant — patients with complex social needs.

A third pattern recurs in AI deployments for patient matching. Many AI workflows require linking records across systems—for example, matching a patient in the EHR to one in the laboratory system or the claims database. If patient matching fails — due to name variations, date-of-birth discrepancies, or missing identifiers — the AI system processes incomplete or incorrect records. The errors are often silent: the model runs without error messages, generates predictions based on incorrect data, and mismatches are discovered only through downstream clinical review or audit.

These failure modes share a common root cause: AI was deployed on a data infrastructure that was not designed to support it. The solution is not to lower expectations for AI

— it is to invest in the data infrastructure that AI requires. FHIR is a central element of that infrastructure.

1.3 FHIR as the Missing Link

HL7 FHIR was published as a draft standard in 2014 and has since become the most important interoperability standard in healthcare. Its adoption is now mandated by U.S. federal regulation for certified EHR systems, required for payer data sharing under CMS rules, and embedded in the API strategies of every major healthcare technology platform. Understanding what FHIR does — and what it does not do — is foundational knowledge for any manager navigating healthcare data strategy.

1.3.1 How FHIR Standardizes and Structures Healthcare Data

FHIR's core innovation is the Resource concept. A Resource is a discrete unit of healthcare data — a Patient, a Condition, a MedicationRequest, an Observation, an Encounter — defined with a standard schema, standard field names, and standard terminology bindings. There are over 150 Resource types in the FHIR R4 specification, covering clinical, administrative, financial, and infrastructure domains. Each Resource has a defined structure: required fields, optional fields, data types, and references to other Resources.

When a laboratory result is represented as a FHIR Observation resource, it will always include a subject (the patient), a code (from LOINC, a standard laboratory terminology), a value (numerical or coded), a unit (from

UCUM, a standard unit system), a status, and an effective date. Any system that understands the FHIR Observation specification — publicly available and freely licensed — can parse this data without additional mapping. Compare this to an HL7 v2 ORU^R01 message, where the same laboratory result might appear in any of several dozen optional fields, encoded in a proprietary code system, with units expressed in free text. The standardization FHIR provides is not cosmetic — it eliminates entire categories of integration work.

FHIR also mandates the use of standard clinical terminologies. LOINC codes for laboratory and clinical observations. SNOMED CT for clinical findings and diagnoses. RxNorm for medications. ICD-10-CM for billing diagnoses. NDC codes for drug products. These terminology bindings create a common semantic layer that allows data from different systems to be compared and aggregated without manual crosswalk development. When a patient's hemoglobin A1c result from one laboratory has the same LOINC code as the same test from a different laboratory, an AI model can treat them as the same clinical variable — because, semantically, they are.

Implementation Guides extend the base FHIR specification for specific use cases. The U.S. Core Data for Interoperability (USCDI) implementation guide defines the minimum data elements that certified EHRs must expose via FHIR APIs. Payer implementation guides define how claims, coverage, and formulary data are structured. Oncology implementation guides define how cancer diagnoses, treatments, and outcomes are represented. These guides create domain-specific standardization on top of FHIR's general-purpose foundation, progressively reducing the variability that complicates data integration.

1.3.2 Why APIs Matter for Real-Time Intelligence

FHIR's RESTful API design is as important as its data model. A RESTful API is a standardized way to request and receive data over HTTP — the same protocol that powers the web. FHIR APIs support standard HTTP operations: GET to retrieve a resource, POST to create one, PUT to update one, and DELETE to remove one. They support query parameters that allow callers to filter resources by patient, date, code, status, and dozens of other criteria. They return responses in JSON or XML, widely supported by programming tools across every major language and platform.

The operational significance of this design is substantial. A FHIR API query to retrieve all of a patient's active medications takes a single HTTP GET request and returns a FHIR Bundle of MedicationRequest resources within seconds. The same operation against a legacy HL7 v2 interface requires establishing a connection, sending a query message, parsing the response segments, and handling the proprietary extensions. The difference in development time between these two approaches is measured in weeks, not hours. Across an enterprise with dozens of integration points, this compounds into a significant reduction in integration cost and complexity.

For AI systems specifically, real-time API access enables use cases that are categorically impossible with batch data pipelines. A clinical decision support model that needs to assess a patient's current risk level at the time of an ordering event cannot wait for a nightly ETL job. A medication safety system that checks for interactions against the patient's current medication list must retrieve that list at the time of the interaction, not from a snapshot taken 24 hours earlier. FHIR APIs make these real-time retrieval patterns technically feasible and practically achievable without custom infrastructure for each use case.

The SMART on FHIR framework, which combines FHIR APIs with OAuth 2.0 authorization, extends this real-time access to third-party applications with appropriate access controls. A clinical analytics application can request authorization to access specific patient data, receive a scoped access token, and query FHIR APIs within the authorized scope — all without embedding credentials in application code or requiring custom authentication schemes. This has direct implications for AI application deployment: SMART on FHIR provides the security and identity layer that allows AI models to operate in production clinical environments with the access controls that regulatory and clinical governance frameworks require.

1.3.3 FHIR as the Language AI Can Understand

The metaphor of FHIR as a language is not merely rhetorical. Language, in the computational sense, is a set of symbols with agreed-upon meanings and grammatical rules that allow information to be expressed unambiguously and interpreted correctly by any party that knows the language. FHIR creates exactly this for healthcare data: a vocabulary of Resource types and data elements, a grammar of schema constraints and terminology bindings, and a syntax of JSON or XML serialization.

AI systems, at their core, require inputs that are defined, consistent, and unambiguous. A machine learning model cannot process a concept; it processes a number, a code, or a string. FHIR's insistence on coded values — LOINC for laboratory observations, SNOMED for clinical findings, RxNorm for medications — means that the same clinical concept is always represented by the same code, regardless of which system originated the data. This consistency is what allows AI models to generalize across data from different sources: the features the model learned to associate with clinical outcomes are stable, not artifacts of a particular system's data encoding.

Diagram 1.4 – FHIR as the Common Language for AI Inputs

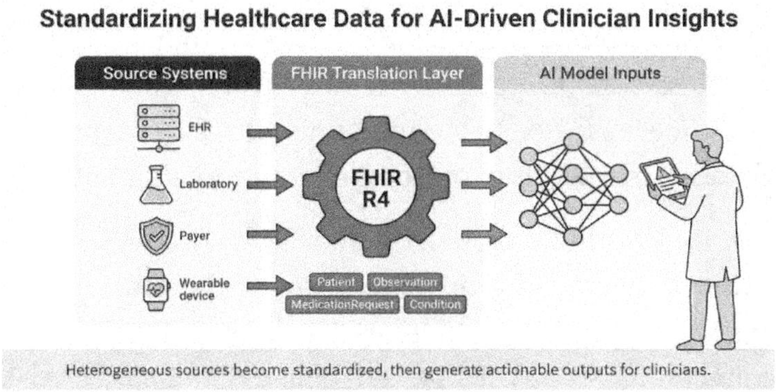

This linguistic consistency also enables transfer learning — a technique in which a model trained on data from one institution is fine-tuned for deployment at another. Transfer learning assumes that the features used during pretraining are semantically equivalent to the features available at the deployment site. Without standardized data encoding, this assumption fails: a model pre-trained on SNOMED-coded diagnoses from one health system cannot be transferred to a site that uses free text or a proprietary coding system without extensive reengineering. FHIR's terminology requirements make transfer learning across institutions feasible in ways that were not previously possible.

For managers evaluating AI vendor claims, the question to ask is: what data format do your models require as input, and what data format do they produce as output? A vendor whose models require custom data feeds that must be built and maintained by the health system imposes significant infrastructure costs and creates technical debt. A vendor whose models operate natively on FHIR-formatted data using standard FHIR APIs is aligned with industry direction and imposes far lower long-term integration costs. This distinction directly affects the pilot-to-production pathway: FHIR-native AI solutions can move from pilot to production faster because the data infrastructure they depend on is already being built to meet regulatory requirements.

It is worth being precise about what FHIR does not do, because overstating its capabilities leads to misaligned expectations. FHIR defines data structures and API conventions — it does not guarantee data quality. An EHR that is technically FHIR-compliant can still have missing data, incorrect codes, and incomplete patient records. FHIR compliance is a necessary but not sufficient condition for AI readiness. The standard creates the infrastructure for data quality improvement; achieving that improvement requires governance, workflow design, and ongoing data stewardship.

1.4 The Symbiotic Relationship

The relationship between FHIR and AI is genuinely symbiotic — not a one-way dependency in which AI consumes FHIR data, but a mutual reinforcement in which each technology enhances the value and capabilities of the other. This framing matters for managers because it has strategic implications: organizations that invest in FHIR infrastructure create not only an interoperability asset but also an AI readiness asset, and vice versa—the two investments compound.

1.4.1 FHIR Enables AI, AI Amplifies the Value of FHIR

The forward direction of the symbiosis is straightforward: FHIR provides the structured, standardized, API-accessible data that AI models need to function reliably in production. As explored in the preceding sections, this is not a theoretical benefit — it is an operational requirement. Without FHIR (or an equivalent standardization layer), AI initiatives in healthcare face data preparation costs that crowd out investment in model development and deployment. FHIR reduces those costs by orders of magnitude and enables the real-time data access that the highest-value AI use cases require.

The reverse direction is equally important and less frequently articulated. AI amplifies the value of FHIR by transforming standardized data into actionable intelligence. A FHIR Observation resource containing a patient's lab result has informational value — it is accessible, queryable, and machine-readable. That same resource, when processed by a machine learning model alongside hundreds of other FHIR resources representing that patient's clinical history, generates predictive value: a probability of deterioration, a recommendation for intervention, an alert that a clinical guideline has not been followed. The FHIR resource has always been there; AI converts it from a data record into a decision-support signal.

This amplification effect compounds over time. As AI systems analyze FHIR-formatted data at scale, they surface patterns that would otherwise be invisible: population-level trends, rare adverse drug events, subtle clinical deterioration pathways, and social determinants that predict care utilization. These insights, fed back into clinical workflows and operational planning, create demand for even more comprehensive and accurate FHIR data. Health systems that recognize the operational value of AI-derived insights are more likely to invest in improving data completeness and the quality of FHIR implementation, which, in turn, improves AI model performance. The loop is self-reinforcing.

Diagram 1.5 – The FHIR-AI Symbiosis Loop

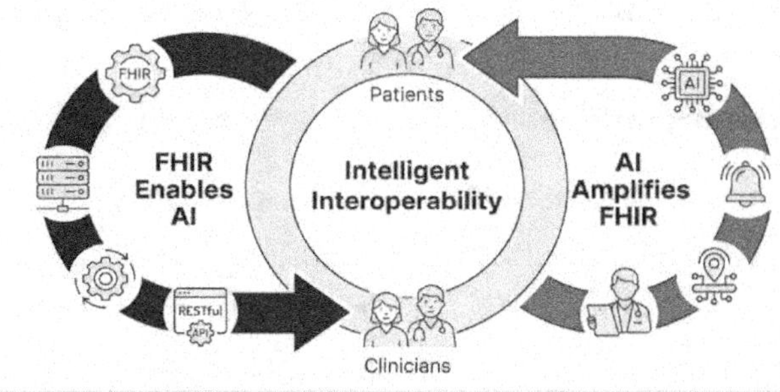

Virtuous cycle help executive and clinicians

From a financial perspective, this compounding means that investments in FHIR infrastructure and investments in AI capability should be evaluated together, not separately. A health system that evaluates its FHIR implementation purely as a regulatory compliance cost is missing the strategic asset it is creating. Equally, a health system that evaluates AI initiatives without accounting for FHIR infrastructure readiness is underestimating both the cost and the timeline of its AI roadmap. Mission-aligned strategy requires treating FHIR and AI as a joint platform investment, not as separate line items.

1.4.2 Intelligent Interoperability as the Next Frontier

Interoperability, as traditionally defined, means that two systems can exchange data in a mutually intelligible format. A patient's medication list sent from a hospital EHR to a community pharmacy system, received and parsed correctly — that is basic interoperability. It is necessary, valuable, and still not fully achieved across the U.S. healthcare system, despite decades of effort. But it is also not sufficient for the demands of modern healthcare. Basic interoperability moves data from point A to point B. Intelligent interoperability moves data from point A to point B and generates insight along the way.

Consider a patient being discharged from a hospital after a myocardial infarction. Basic interoperability means the discharge summary is electronically transmitted to the patient's primary care physician. Intelligent interoperability means that as the discharge summary is generated, an AI model identifies that the patient has not been prescribed a statin, fires a recommendation to the discharging physician, and simultaneously enrolls the patient in a post-discharge monitoring program that will push FHIR-formatted vital sign readings from the patient's home blood pressure cuff to the care team's dashboard. The medication gets prescribed. The care gap gets closed. The readmission risk decreases.

This is not a speculative future scenario. Each component of this workflow — FHIR-based discharge summary exchange, AI-driven clinical decision support, SMART on FHIR application launch, remote patient monitoring via FHIR Observations — is technically feasible today using commercially available or open-source tools. What prevents this workflow from being standard practice is not technology immaturity; it is the uneven deployment of FHIR infrastructure, the fragmented governance of AI applications, and the organizational and contractual barriers that prevent data sharing across institutional boundaries.

For managers, intelligent interoperability represents a concrete operational goal — not an aspiration. The question is not whether to pursue it but how to sequence the investments required to get there. That sequencing is the subject of subsequent chapters. The foundational principle established here is that intelligent interoperability requires both FHIR and AI to operate together, and that neither alone is sufficient.

1.4.3 Why Organizations That Embrace Both Will Lead the Future

Healthcare organizations that have invested in FHIR infrastructure and AI capabilities simultaneously are beginning to demonstrate measurable competitive and clinical advantages. These advantages manifest across several dimensions that healthcare executives monitor closely.

Clinical quality outcomes are improving in organizations that deploy AI-powered decision support on real-time FHIR data. Sepsis mortality rates, medication reconciliation accuracy, preventive care gaps, and early identification of patients at risk for deterioration are all amenable to AI-driven improvement — but only when the underlying data is accessible, complete, and standardized. Organizations that have invested in FHIR data quality see their AI use cases perform closer to the published benchmarks from academic research; organizations that have not see a persistent performance gap between research claims and production results.

Operational efficiency gains are equally significant. AI-assisted prior authorization, automated referral routing, intelligent documentation, and predictive staffing models all reduce administrative burden and improve resource allocation. These gains are not hypothetical — they are documented in post-implementation analyses from health systems that have moved AI use cases from pilot to production. The pilot-to-production pathway is consistently faster at organizations with mature FHIR infrastructure, because the data access and integration work that consumes months in immature environments is already done.

Market position is a third dimension. As patients, payers, and regulators increasingly demand transparent, data-driven, interoperable care, health systems that cannot demonstrate these capabilities will face growing disadvantages in contracting, patient acquisition, and regulatory standing. The information-blocking rules under the 21st Century Cures Act impose real financial penalties on organizations that fail to enable appropriate data access. CMS value-based care programs increasingly require data submission in FHIR formats. Health systems that have invested in FHIR as a strategic asset are better positioned to participate in these programs and to negotiate from a position of strength in payer and partner relationships.

The organizations that will lead the next decade of healthcare are those that treat FHIR and AI not as separate technology initiatives but as a unified platform strategy. This means data governance frameworks that prioritize FHIR data quality. It means AI governance frameworks that enforce responsible-by-design principles — including explainability, bias monitoring, and clinical validation. It means organizational alignment between clinical informatics, IT, legal and compliance, and clinical operations, because the intersection of FHIR and AI touches all of these domains simultaneously.

Diagram 1.6 – The Strategic Advantage Framework: FHIR + AI Maturity

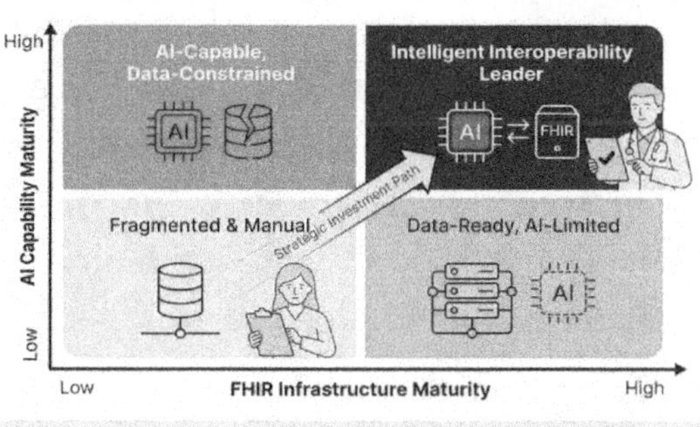

The decision architecture for any healthcare manager reading this book can be summarized as: identify where your organization sits on the FHIR-AI maturity spectrum, understand the specific gaps constraining your current position, and sequence your investments to close those gaps in an order that generates operational value at each step. That sequencing — the pilot-to-production pathway for FHIR-enabled AI — is the practical subject this book is designed to support.

1.5 Takeaway

The healthcare data problem is real, measurable, and expensive. Fragmentation across EHRs, payers, laboratories, and devices produces clinical records that are systematically incomplete at the point of care. Legacy standards were not designed to support real-time, machine-readable access to data. The cost of this inadequacy manifests in denied claims, duplicate tests, medication errors, clinician burnout, and failed AI deployments.

AI in healthcare has matured from rule-based alerting into a broad toolkit of machine learning and natural language processing capabilities. Those capabilities are real. But their production performance is bound by data quality. Models trained on clean, standardized data and deployed against consistent, real-time data feeds perform as advertised. Models deployed on fragmented, inconsistent, batch-processed data perform at a fraction of their potential — and erode clinician trust in the process.

FHIR is the standard that closes the gap between the data healthcare organizations have and the data AI requires. It defines resources with a consistent structure and standard terminology. It mandates RESTful APIs that enable real-time access to data. It supports the SMART on FHIR authorization framework that allows AI applications to operate securely in production environments. FHIR does not guarantee data quality — but it creates the infrastructure for achieving and maintaining it.

The relationship between FHIR and AI is not a dependency; it is a symbiosis. FHIR enables AI to function reliably in clinical production. AI amplifies the operational value of FHIR by transforming standardized data into actionable intelligence. Together, they create intelligent interoperability — the capacity of healthcare systems not just to exchange data, but to learn from it, act on it, and improve continuously.

Organizations that invest in this joint platform — FHIR infrastructure and AI capability, governed together, deployed responsibly — are building the operational foundation for mission-aligned healthcare delivery in the decade ahead. Those who treat them as separate initiatives or defer investment until regulatory pressure forces their hand will find themselves behind in clinical quality, operational efficiency, and market position.

The key decisions this chapter informs: Evaluate your current FHIR implementation maturity honestly, accounting for data completeness and API accessibility — not just regulatory compliance checkboxes. Assess AI initiatives for FHIR-native data requirements before committing to vendor contracts. Treat FHIR infrastructure investment and AI investment as a unified platform strategy. And establish the data governance foundation — terminology standards, patient matching, access controls — before deploying AI in production clinical workflows.

2 Primer on FHIR: The Standard That Changed Everything

Every major shift in healthcare IT has come down to the same problem: getting systems to talk to each other without requiring armies of analysts, custom code, and years of maintenance. For decades, the industry tried to solve this with well-intentioned but fundamentally misaligned standards for how modern software works. Then FHIR arrived. Fast Healthcare Interoperability Resources — FHIR, pronounced "fire" — is the standard that finally aligned healthcare data exchange with the web-native world that every other industry had already embraced.

This chapter is not a technical deep dive for developers. It is a working primer for managers, executives, and IT leaders who need to understand what FHIR is, why it matters operationally, and how it enables AI to function at scale in healthcare. If you cannot explain FHIR to a board member in two minutes, or if you are uncertain what questions to ask your EHR vendor about FHIR compliance, this chapter fills that gap. The decisions you make about FHIR adoption today will determine whether your organization has clean, usable data for AI tomorrow.

FHIR is not just a technical upgrade. It is an architectural shift. It changes how data is modeled, how systems query each other, how patients access their own records, and how AI models are trained and deployed. Understanding FHIR is now a management competency, not just an IT specialty.

2.1 Why FHIR Was Created

2.1.1 Limitations of HL7 v2, CDA, and Custom Interfaces

To understand why FHIR matters, you need to understand what came before it — and why those predecessors failed to scale. The healthcare industry has been attempting to standardize data exchange since the mid-1980s. Health Level Seven International (HL7) released its version 2 messaging standard in 1989. HL7 v2 became the most widely deployed health IT standard worldwide. It is also one of the most fragmented.

HL7 v2 messages are pipe-delimited text strings — a series of segments like MSH, PID, PV1, and OBX — transmitted over MLLP, a low-level TCP protocol. The standard was designed for a world of mainframes and point-to-point hospital connections. It predates the modern web, RESTful APIs, JSON, and cloud computing by decades. The core problem with HL7 v2 is not that it was poorly designed — it was well-designed for its era. The problem is that the standard left too much to local interpretation. A "message" sent by one vendor's system and received by another rarely arrived in the same shape it was sent. Hospitals routinely built custom interface engines to translate between two HL7 v2 dialects.

The workflow-level impact of this fragmentation is enormous. Interface engines — software platforms like Mirth Connect, Rhapsody, or Cloverleaf — are required to translate, transform, and route messages between systems. Every new connection between two systems typically requires a custom interface build: weeks of analysis, development, testing, and ongoing maintenance. A mid-sized hospital may run hundreds of active interfaces. A large health system can have thousands. Each interface is a potential failure point, a maintenance burden, and a cost center.

HL7 v3, released in the early 2000s, attempted to address the ambiguity problem by introducing a rigid, XML-based model grounded in a formal information architecture known as the Reference Information Model (RIM). The theory was sound: a unified data model would eliminate the variation that plagued v2. In practice, HL7 v3 was so complex that adoption was minimal. Implementation guides ran to hundreds of pages. Tooling was limited. Most vendors declined to implement it.

The Clinical Document Architecture, or CDA, emerged as a more pragmatic subset of HL7 v3. CDA is still widely used today, particularly for clinical summaries, referral letters, and care transition documents. The Consolidated CDA (C-CDA) became the federally mandated format for Meaningful Use Stage 2 in the United States. CDA documents solved the document-exchange problem — sending a discharge summary from one hospital to another — but they were not designed for granular, query-based data access. Parsing a C-CDA to extract a single lab result requires navigating a large XML document with complex namespaces and sections. For AI pipelines that need structured, atomic data, C-CDA is a poor source.

The net result of this history is an industry built on point-to-point custom interfaces, a patchwork of partially overlapping standards, and data silos that resist integration. For a manager, the practical consequences are predictable: high integration costs, long implementation timelines, data quality problems rooted in inconsistent formatting, and AI projects blocked waiting for clean data. The case for FHIR begins here.

Diagram 2.1 – The Pre-FHIR Integration Landscape

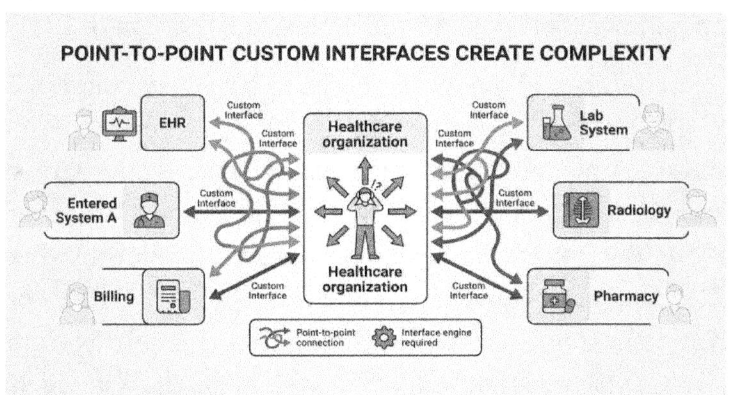

2.1.2 The Need for Modern, Web-Native Interoperability

By the late 2000s, the contrast between healthcare IT and the rest of the technology world was stark. Consumer apps, financial services platforms, and e-commerce systems had converged on a set of architectural principles that made integration straightforward: REST APIs, JSON payloads, OAuth for authorization, and HTTP as the universal transport. A developer could connect to a bank's API, a social platform, or a payment processor in hours. Connecting two EHR systems required months.

Grahame Grieve, an Australian healthcare informaticist, began articulating an alternative approach around 2011 under the working name "Resources for Healthcare." The idea was to abandon the complexity of HL7 v3 and CDA in favor of a model that mirrored how the modern web actually worked. Data would be broken into discrete, well-defined units — resources — each with a standard structure, a stable URL, and the ability to be accessed via standard HTTP operations. This was the conceptual foundation of FHIR.

HL7 formally adopted FHIR as a draft standard in 2012. FHIR Release 4, published in 2019, achieved normative status — meaning core parts of the specification are stable enough for production use without risk of breaking changes. FHIR R4 is now the baseline standard required by the U.S. 21st Century Cures Act, enforced by the Office of the National Coordinator for Health Information Technology (ONC). Every certified EHR system in the United States must expose an FHIR R4 API. This regulatory mandate, more than any technical argument, is why FHIR adoption accelerated dramatically after 2020.

From a manager's perspective, the shift to web-native interoperability means that the cost of building new integrations drops significantly when both sides speak FHIR. Instead of building a custom interface engine, developers can query a FHIR server with standard HTTP GET requests, receive a JSON response, and map the data into their application. The same patterns used to build any web application apply directly. This dramatically expands the pool of developers who can work on healthcare integrations and compresses implementation timelines from months to weeks — or even days for straightforward use cases.

For AI teams, the implications are equally significant. Machine learning pipelines need data at scale, in consistent formats, available programmatically. FHIR APIs enable automated data pipelines to query patient records, extract specific resources, and feed structured data into training and inference workflows without manual extraction or custom parsing. The pilot-to-production pathway for AI in healthcare is substantially shorter when FHIR is the data layer.

2.2 Core Concepts

2.2.1 Resources: The Building Blocks

A FHIR resource is the atomic unit of healthcare information. Think of it as a structured, self-contained data object that represents one clinical or administrative concept. HL7 has defined over 150 resource types in FHIR R4. Each resource has a defined set of fields, a JSON or XML serialization format, a unique identifier, and a URL-based address. Resources are designed to be both human-readable and machine-processable.

The most commonly used resources in clinical workflows include Patient, Encounter, Condition, Observation, MedicationRequest, DiagnosticReport, Procedure, AllergyIntolerance, CarePlan, and Practitioner. Each of these maps directly relates to a real concept that clinicians, administrators, and IT teams already understand. A Patient resource contains demographic information. An

Observation resource contains a single lab result or vital sign. A MedicationRequest contains a medication order. The resource model is intuitive precisely because it mirrors clinical reality.

Each resource instance is identified by a URL that follows a predictable pattern: the base URL of the FHIR server, the resource type, and a unique ID. For example, a patient record might live at https://fhir.examplehospital.org/Patient/12345. This addressability is what makes FHIR resources so powerful for integration: any authorized system can retrieve a specific resource — or query for a set of resources matching certain criteria — using standard HTTP.

Resources reference each other using these URLs. An Encounter resource references the Patient to which it belongs. A MedicationRequest references the Encounter, the Patient, and the Practitioner who ordered it. This linked graph of resources makes it possible to reconstruct a complete clinical picture from a FHIR server without requiring bespoke database joins or custom extraction logic. For AI applications, this graph structure is particularly valuable: it makes it straightforward to programmatically assemble a longitudinal patient record.

From a manager's standpoint, understanding resources means understanding what your FHIR implementation covers. Not every vendor exposes every resource type. An EHR vendor might fully implement Patient, Encounter, Condition, and Observation, but provide limited support for CarePlan or Provenance. When evaluating FHIR compliance, the right question is not just "do you support

FHIR R4?" but "which resources do you expose, at what conformance level, and what data elements are populated?" Operational clarity on this point is essential before committing AI use cases to a FHIR data source.

Diagram 2.2 – FHIR Resource Graph for a Patient Encounter

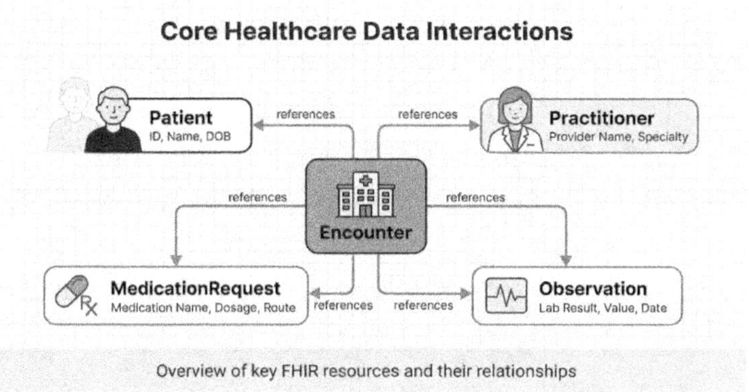

Overview of key FHIR resources and their relationships

2.2.2 Profiles and Extensions

The base FHIR specification defines resources with a broad set of optional fields designed to accommodate the full range of global healthcare scenarios. In practice, different use cases, countries, and organizations need to constrain or extend these base definitions. This is where profiles and extensions come in.

A FHIR profile is a formal constraint on a base resource. It specifies which elements are required (cardinality), which value sets are valid for coded elements, and which extensions are expected. Profiles are expressed using the StructureDefinition resource. The US Core Implementation Guide, for example, defines a US Core Patient profile that mandates certain demographic elements and constrains the identifier types to those commonly used in the United States. When a system claims to be "US Core compliant," it means its resources conform to the US Core profiles.

Extensions allow implementers to add data elements not present in the base resource definition. Healthcare is endlessly varied, and no specification can anticipate every local requirement. A state health department might need to capture a patient's tribal affiliation. A specialty clinic might need to track a proprietary risk score. Extensions provide a structured, schema-defined way to add these elements without breaking compatibility with systems that do not understand them. A system that receives a resource with an unrecognized extension is expected to ignore it gracefully — a principle FHIR calls "mustSupport."

For managers, profiles are the governance mechanism that makes FHIR interoperability real rather than theoretical. Two systems that both "support FHIR R4" may not be interoperable if they implement different profiles with conflicting constraints. Before declaring interoperability, teams must verify profile alignment — that both systems use the same implementation guide. This is a governance and procurement decision as much as a technical one. Implementation guides like US Core, Da Vinci (for payer-provider exchange), and USCDI+ are the standards on which interoperability commitments should be based.

Managing profiles is also a data quality issue directly relevant to AI. If your AI model is trained on data from systems using inconsistent profiles — one that populates a field as required and another that treats it as optional — the resulting dataset will have systematic gaps. Mission-aligned AI programs invest in profile governance as part of their data strategy, not as an afterthought.

2.2.3 RESTful APIs and Standard Operations

FHIR's most operationally significant design choice is its use of RESTful APIs as the primary interaction model. REST — Representational State Transfer — is the architectural style underlying the modern web. Every time a browser loads a webpage, it is making a REST request. Every major consumer platform — from Google Maps to banking apps to payment processors — exposes REST APIs. FHIR brings this model to healthcare data.

The FHIR RESTful API defines a small set of standard interactions that apply uniformly across all resource types. These interactions map directly to standard HTTP methods. The core operations are read (HTTP GET for a specific resource), search (HTTP GET with query parameters), create (HTTP POST), update (HTTP PUT), and delete (HTTP DELETE). A capability statement — the CapabilityStatement resource — declares which operations a FHIR server supports, making it self-documenting.

Search is the interaction most relevant to AI use cases. FHIR search allows systems to query for resources based on specific criteria using URL query parameters. A query like GET /Observation?patient=12345&code=2339-0&date=ge2024-01-01 retrieves all blood glucose observations for patient 12345 on or after January 1, 2024. These search parameters are standardized — you do not need to know the internal schema of the EHR database to write this query. The FHIR server translates it to whatever internal representation it uses.

FHIR also supports Bulk Data Access — defined in the SMART Backend Services specification — which allows authorized systems to export large datasets asynchronously. Rather than querying records one by one, a bulk export retrieves all resources of specified types for a population. This is the pattern used for population health analytics, AI training data pipelines, and public health reporting. For any AI use case that requires training data from thousands of patients, bulk FHIR export is the production-grade approach.

The operational clarity that RESTful FHIR APIs provide over legacy interfaces is substantial. Legacy HL7 v2 interfaces require monitoring, error queuing, custom acknowledgment logic, and interface engine infrastructure. A FHIR API call, by contrast, returns a standard HTTP response with a standard status code. Errors are described in an OperationOutcome resource. Developers can test FHIR APIs with standard web tools such as curl, Postman, or any HTTP client, without specialized healthcare middleware. This lowers the barrier to entry for new development teams and shortens the pilot-to-production pathway for new applications.

2.2.4 JSON vs XML in FHIR

FHIR resources can be serialized in multiple formats: JSON, XML, and a compact format called FHIR Shorthand (FSH), used primarily for authoring implementation guides. In practice, the choice for most modern implementations is JSON. JSON is lighter weight than XML, easier to read and write, natively supported by JavaScript and most modern programming languages, and the dominant format for REST API payloads across the software industry.

XML remains relevant in specific contexts. Some legacy systems and older implementation guides use XML as the default. C-CDA documents are XML-based, and some workflows bridge between CDA and FHIR. Certain FHIR operations — particularly those involving XSLT-based transformations in legacy pipeline architectures — continue to use XML. But for net-new implementations, API-based integrations, and AI data pipelines, JSON is the practical standard.

For managers evaluating technology choices, the JSON versus XML question has vendor selection implications. Confirm that your EHR vendor's FHIR API returns JSON by default, and that your data engineering team's tools — pipeline frameworks, data lakes, ML platforms — can ingest FHIR JSON natively or with minimal transformation. Most modern data platforms handle JSON with no special configuration. XML requires additional parsing libraries and is more verbose, making it less efficient for high-volume data pipelines.

The JSON representation of a FHIR resource is structured as a nested object with a resourceType field at the top level, followed by the resource's defined elements. A Patient resource in JSON includes fields like id, identifier, name, gender, birthDate, and address. Each element follows a consistent naming convention and data type. This consistency is what allows AI teams to write generic parsers that work across resource types rather than building custom extraction logic for every data source.

Diagram 2.3 – FHIR RESTful API Interaction Model

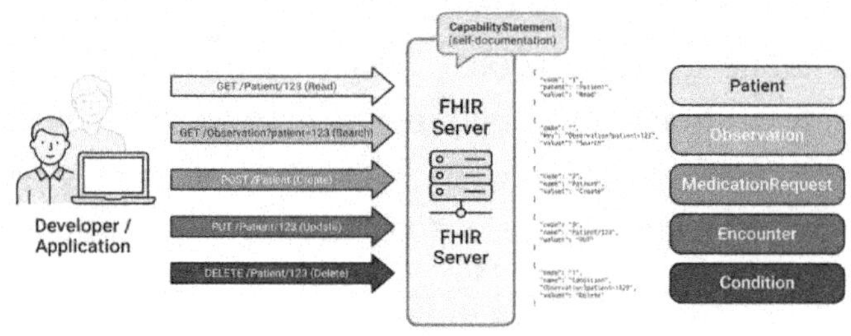

2.3 How FHIR Improves Data Quality

2.3.1 Consistency Across Systems

Data quality is the single most important factor in whether AI delivers value in healthcare. Bad data does not produce wrong answers — it produces confidently wrong answers, which is worse. The fundamental data quality problem in healthcare has always been inconsistency: the same clinical concept represented in dozens of different ways across different systems, each with its own data model, code sets, and naming conventions.

Consider how a patient's list of problems is stored across a typical health system. The inpatient EHR uses ICD-10 codes mapped to a proprietary problem list model. The ambulatory EHR uses SNOMED CT codes with a different status field structure. The specialty system uses free-text descriptions of problems. The payer's claims data uses ICD-10 diagnosis codes tied to encounter claims but lacks the temporal and severity context in the clinical systems. An analytics team trying to identify all patients with a specific condition must reconcile four different representations of the same clinical concept. This is not an unusual scenario — it is the norm.

FHIR addresses this through two mechanisms: resource standardization and terminology binding. The Condition resource defines a structured representation of a clinical problem or diagnosis. The code element is bound to standard terminologies — ICD-10, SNOMED CT, or ICD-9 — with explicit value set constraints. The clinicalStatus and verificationStatus elements use standardized code systems. The onset and abatement elements follow standard date formats. When systems comply with the same FHIR profile for the Condition resource, data from those systems can be combined without manual reconciliation.

This does not mean FHIR eliminates all data quality problems. Profile compliance varies. Terminology mapping is imperfect. Some systems populate only mandatory fields and leave optional elements blank. But the direction of travel is clear: FHIR defines a standard target state, and the regulatory requirements around FHIR compliance create market pressure on vendors to conform. Over time, the consistency gains compound. An organization that insists on strong FHIR profile conformance in its vendor contracts is investing in progressively cleaner data for AI.

Responsible by design means building these quality standards into procurement, integration, and governance processes — not trying to clean up inconsistent data after the fact. The cost of fixing a data quality problem in a running AI system is an order of magnitude higher than preventing it at the source. FHIR profile conformance is a preventive measure.

2.3.2 Cleaner Data for Analytics and AI

Beyond structural consistency, FHIR improves data quality for analytics and AI in several concrete ways. First, FHIR resources carry provenance — information about where data came from, who recorded it, and when. The Provenance resource links data elements to their origin system, recorder, and timestamp. For AI models, provenance is critical: a lab result from a certified reference laboratory has different reliability characteristics than a patient-reported value. Models that use provenance metadata can appropriately weight these sources.

Second, FHIR's use of standard terminologies reduces the normalization burden on AI teams. When observations use standard LOINC codes, medications use RxNorm, and diagnoses use ICD-10 or SNOMED CT, data from different sources are already in a common vocabulary. Feature engineering — the process of preparing data for model training — is substantially simpler. A LOINC-coded lab result from System A and the same LOINC-coded result from System B can be treated as the same feature without manual mapping.

Third, FHIR's atomic resource model supports granular data access. Instead of receiving a dense C-CDA document and parsing out the relevant elements, an AI pipeline can request exactly the resource types it needs — for example, all Observations of type vital sign for a cohort of patients over a specific date range. This targeted access reduces noise in training data and makes it easier to maintain reproducible data pipelines.

Fourth, FHIR's query model supports cohort identification — the process of selecting the patient population for an AI model. Researchers can query FHIR APIs with specific clinical criteria (conditions, medications, age ranges, encounter types) to build study cohorts programmatically. This makes it possible to design and execute studies that would previously have required manual chart review or expensive data extraction projects. The practical effect is faster iteration — more hypotheses tested, more models validated, in less time.

The combination of these factors — provenance, standard terminologies, atomic access, and cohort query support — means that FHIR data is significantly more tractable for AI than the fragmented, heterogeneous data landscape that preceded it. It does not solve every problem. But it raises the floor of data quality to a level where AI projects can reach production rather than stalling in the data preparation phase.

2.3.3 Real-World Examples of FHIR Simplifying Integration

Abstract benefits become meaningful when grounded in operational examples. The following cases illustrate how FHIR has changed the integration calculus in real healthcare environments.

Example 1: Patient-Matching for Care Coordination

A large regional health system was operating three separate EHR instances following a series of acquisitions. Care coordinators needed to see a unified patient record spanning all three systems to manage high-risk patients effectively. The legacy approach — a custom master patient index with point-to-point HL7 v2 feeds from each EHR — required eighteen months of interface development and produced a record that was hours out of date due to batch synchronization. When the health system migrated all three EHRs to FHIR R4 APIs, a care coordination application could query all three FHIR endpoints in real time using FHIR Patient search and $match operations. The unified view was available in near-real time after six weeks of development — a fraction of the original timeline and cost.

Example 2: AI-Powered Readmission Risk

A hospital system deployed a readmission risk model that needed daily updates with the latest clinical data for recently discharged patients. Under the prior architecture, this required a nightly extract from the EHR database, a complex ETL process, and manual quality checks — a pipeline that regularly failed and took up to four hours to complete. With FHIR Bulk Data export, the same data was available via an asynchronous export job that completed in under thirty minutes and produced structured NDJSON files directly consumable by the model's feature engineering pipeline. Operational staff no longer needed to monitor and restart a fragile custom extract process.

Example 3: Medication Reconciliation Across Transitions of Care

Medication reconciliation at care transitions — particularly from inpatient to ambulatory settings — is a persistent patient safety problem. Discrepancies between the inpatient medication list and the ambulatory prescription record are common and clinically dangerous. A health system implemented a FHIR-based medication reconciliation tool that queried both the hospital and ambulatory EHRs using standard MedicationRequest and MedicationStatement resources. The tool surfaced discrepancies to a pharmacist within seconds of discharge documentation. Previously, pharmacists performed this reconciliation manually by switching between two systems and comparing printed medication lists — a process that took fifteen to twenty minutes per patient and was frequently skipped under time pressure.

Diagram 2.4 – FHIR Data Flow from Clinical Systems to AI Pipeline

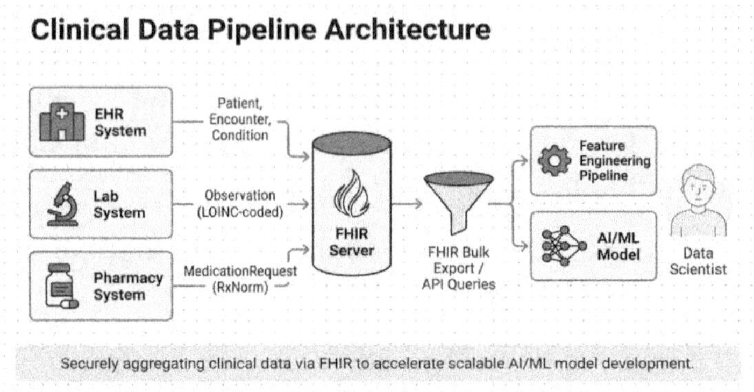

2.4 FHIR in the Real World

FHIR's value is not evenly distributed across all healthcare contexts. It plays out differently depending on whether the primary actor is a provider, a payer, a patient-facing application developer, or a public health agency. Understanding these different deployment contexts helps managers set realistic expectations and allocate resources appropriately.

2.4.1 Provider Workflows

For provider organizations — hospitals, health systems, physician practices, and specialty clinics — FHIR's most immediate impact is on clinical decision support and workflow integration. The SMART on FHIR application framework, built on top of the FHIR API and OAuth 2.0, allows third-party applications to embed within EHR workflows with contextual access to the patient's data. A SMART on FHIR application launches inside the EHR's interface, receives the current patient context, queries the

FHIR API for relevant data, and presents clinical guidance without requiring the clinician to leave the EHR.

This architectural pattern has significant operational implications. Before SMART on FHIR, integrating a third-party clinical decision support tool into an EHR workflow required vendor cooperation, custom API development, and typically a lengthy certification process. SMART on FHIR standardizes this integration pattern so that any compliant application can integrate with any compliant EHR. A sepsis detection tool, a drug-drug interaction checker, or an AI-powered diagnostic support application built to the SMART on FHIR standard can, in principle, run on multiple EHR platforms without per-platform redevelopment.

For hospital IT departments, FHIR changes the economics of application integration. Rather than negotiating bespoke API access with each EHR vendor for each new application, teams can evaluate SMART on FHIR applications against a standardized checklist: FHIR R4 conformance, US Core profile support, OAuth 2.0 authorization, and appropriate data scopes. This creates operational clarity in vendor evaluation and procurement. It also reduces the risk of vendor lock-in, since applications built to open standards are more portable.

In provider workflows, the most common FHIR resource types in active use are Patient (demographics and identifiers), Encounter (visit records), Condition (problem list and diagnoses), Observation (lab results, vital signs, clinical assessments), MedicationRequest (orders), DocumentReference (clinical notes and documents), and AllergyIntolerance. These resources collectively represent the core of a patient's longitudinal clinical record and form the foundation for most clinical AI applications.

Workflow-level impact for provider organizations also extends to care coordination across institutional boundaries. FHIR-enabled patient record queries at the point of care — retrieving records from external facilities when a patient presents — can surface critical information that would otherwise be unknown to the treating team. Prior authorizations, specialist consult notes, recent emergency department visits at competing facilities, and external lab results can now flow into the EHR in real time rather than arriving by fax days later. The clinical and operational value of this capability is substantial.

2.4.2 Payer Data Exchange

Health insurers and payers operate at the intersection of clinical and administrative data — and historically, they have been among the least interoperable actors in the healthcare system. Claims data is rich in administrative detail but sparse in clinical context. Clinical data from provider systems was largely inaccessible to payers, creating gaps in care management, prior authorization, and risk adjustment processes.

The CMS Interoperability and Patient Access Rule, finalized in 2020 and significantly expanded through subsequent rulemaking, mandated that Medicare Advantage, Medicaid, and CHIP payers expose FHIR R4 APIs for member data access. This was a watershed moment: for the first time, federal regulation required payers to make member data available programmatically through a standards-based API. Members gained the right to direct their data to third-party applications. Payers were required to build and maintain the infrastructure to support this access.

The Da Vinci Project, an HL7 FHIR Accelerator program with participation from major payers and providers, has developed a suite of implementation guides tailored to payer use cases. These include Prior Authorization Support (PAS), Coverage Requirements Discovery (CRD), Documentation Templates and Rules (DTR), and Payer Data Exchange (PDex). Each addresses a specific pain point in the payer-provider relationship with a standardized FHIR-based workflow.

Prior authorization is one of the most operationally significant use cases for payers. The traditional prior authorization process is manual, phone- and fax-based, and consumes enormous administrative resources on both the payer and provider sides. Studies have estimated that the prior authorization burden costs the U.S. healthcare system billions of dollars annually in administrative overhead and contributes to treatment delays that affect patient outcomes. FHIR-based prior authorization — where a provider's EHR can query a payer's FHIR server for coverage requirements and submit authorization requests electronically — directly addresses this burden.

For payer organizations, FHIR also enables member-to-member data portability during plan transitions. When a member switches from one plan to another, the PDex implementation guide allows the new payer to request the member's claims history from the prior payer via a FHIR API with the member's consent. This capability supports continuity of care, reduces the risk of duplicate testing, and improves the accuracy of risk adjustment calculations — with direct financial implications for the receiving payer.

From an AI perspective, payer access to FHIR clinical data unlocks use cases that were previously limited by data availability. Care management programs that identify high-risk members can draw on richer clinical data — not just claims — to improve predictive accuracy. Population health models can incorporate real-time clinical events rather than lagged claims data. Fraud detection systems can more effectively cross-reference clinical and administrative data. These capabilities have mission-aligned financial and quality outcomes for payer organizations.

Diagram 2.5 – Payer-Provider FHIR Data Exchange (Da Vinci Model)

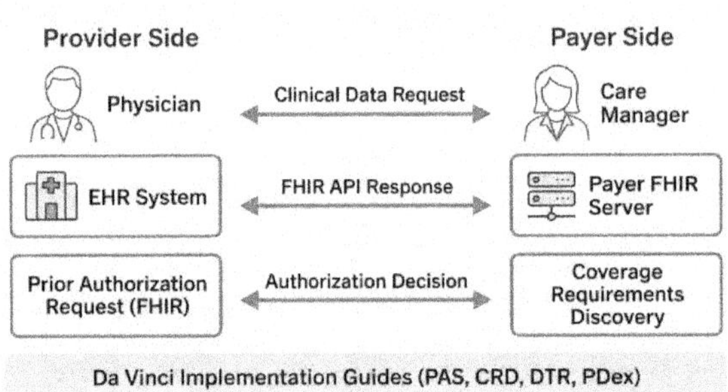

2.4.3 Patient-Facing Apps

The patient-facing dimension of FHIR represents one of the most strategically significant shifts in the history of healthcare data access. The 21st Century Cures Act's information blocking provisions, combined with the ONC and CMS interoperability rules, established that patients have a federal right to access their health information in a standardized, electronic format — specifically via FHIR APIs — and that healthcare organizations cannot obstruct this access without a specified legal exception.

In practical terms, this means that a patient can now direct their EHR data to any application they choose using a standardized OAuth 2.0 and FHIR API flow. They visit their patient portal, authorize a third-party app, and the app receives access to their FHIR data for the duration of the authorized session. Apple Health, for example, implemented FHIR data access through its Health Records feature, allowing patients to pull their clinical data from participating EHRs directly into their iPhone's Health app. This data can then be shared with caregivers, other providers, or health management applications.

For health systems and provider organizations, this represents both an opportunity and a responsibility. The opportunity lies in patient engagement: applications that surface personalized health information, send preventive care reminders, or provide post-discharge instructions using real clinical data tend to outperform generic wellness apps on both engagement and health outcomes. The responsibility lies in patient trust and data stewardship: patients authorize third-party apps based on trust in their healthcare provider's infrastructure. If those apps misuse data, patients attribute the harm to the healthcare organization that enabled access.

The patient-facing FHIR ecosystem also creates new inputs for AI. Patient-reported outcomes, wearable device data, and behavioral data from patient-facing apps can now flow through standardized FHIR interfaces back into clinical systems. The Observation resource, for example, can represent a patient-reported pain score just as readily as a laboratory result. As this bidirectional data flow matures, the richness of the data available to AI models expands significantly — moving beyond the traditional clinical record to incorporate the patient's lived experience between visits.

From a governance standpoint, patient-facing FHIR access requires clear policies on data scope, consent management, and app vetting. Not all third-party apps are equally trustworthy. Organizations that help patients understand which apps are vetted and what data those apps can access are providing a genuine service. Responsible by design means building this guidance into the patient experience, not treating it as a compliance checkbox.

2.4.4 Public Health and Research

Public health agencies and research institutions have long been data consumers rather than data producers in the healthcare system. They depend on data flows from providers, labs, payers, and patients to detect disease outbreaks, monitor population health, and conduct clinical research. The traditional mechanisms for these data flows — case reports, lab reportable conditions, cancer registry submissions, and research data extracts — are largely paper-based, batch-processed, or dependent on custom electronic reporting systems that vary by jurisdiction.

FHIR is transforming this landscape on multiple fronts. The MedMorph (Making Electronic Data More Available for Research and Public Health) initiative, developed by the CDC and ONC, defines a FHIR-based architecture for automated data reporting from clinical care to public health agencies and research networks. Under this model, clinical systems can automatically detect reportable conditions — a positive COVID test, a new cancer diagnosis, an electronic case report — and submit a FHIR-based report to the appropriate agency without manual intervention. The same event-driven architecture that powers real-time EHR notifications to clinical staff can trigger automated public health reporting.

For clinical research, FHIR is enabling a new generation of trial infrastructure. The Clinical Trial protocol and data-sharing infrastructure built on FHIR — including the ResearchStudy and ResearchSubject resources — allows researchers to define eligibility criteria in machine-readable FHIR queries, identify eligible patients across FHIR-enabled clinical systems, and structure data-collection instruments as FHIR Questionnaire resources. The PCORnet Common Data Model and the OMOP Common Data Model both have FHIR mapping specifications, allowing data from FHIR sources to flow into research networks without custom extraction.

For AI researchers, the implications are significant. Research datasets built on FHIR sources are more reproducible: the queries that define the study cohort are explicit, the resource types included are enumerated, and the terminology bindings are documented. This enables external teams to validate and replicate studies across different FHIR-enabled data sources. Reproducibility has been a persistent challenge in healthcare AI, and FHIR-based research infrastructure directly addresses one of its root causes.

Public health surveillance using FHIR also creates opportunities for near-real-time AI applications. Syndromic surveillance systems — which detect patterns suggesting disease outbreaks — traditionally operate on emergency department chief complaints submitted in batches. A FHIR-based surveillance system can receive structured clinical data in near real time, enabling faster outbreak detection with higher clinical specificity. The COVID-19 pandemic made the cost of slow surveillance data painfully apparent; FHIR-enabled infrastructure represents the public health community's investment in avoiding a repeat of that experience.

Diagram 2.6 – FHIR Across the Healthcare Ecosystem

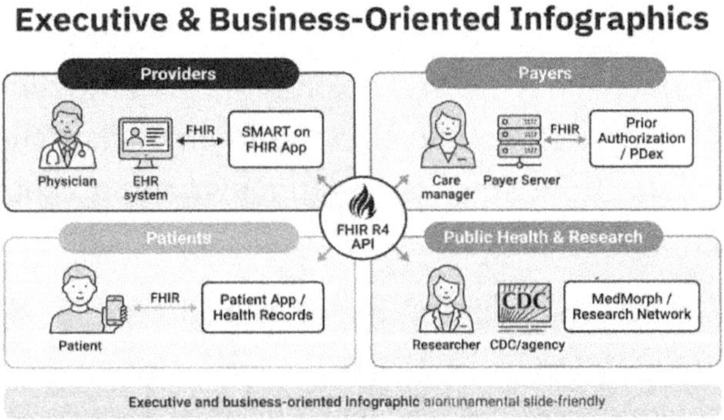

Executive & Business-Oriented Infographics

Providers

Physician — EHR system — FHIR — SMART on FHIR App

Payers

Care manager — Payer Server — FHIR — Prior Authorization / PDex

FHIR R4 API

Patients

Patient — FHIR — Patient App / Health Records

Public Health & Research

Researcher — CDC/agency — CDC — MedMorph / Research Network

Executive and business-oriented infographic aiariunamental slide-friendly

2.5 Takeaway

FHIR is not a vendor product, a consulting engagement, or a one-time migration project. It is a durable architectural standard that is reshaping how every actor in the healthcare system — providers, payers, patients, researchers, and public health agencies — accesses, exchanges, and acts on clinical data. Understanding it at the level of management decision-making is no longer optional.

The pre-FHIR landscape — built on HL7 v2 custom interfaces, CDA documents, and proprietary APIs — was expensive to build, fragile to maintain, and unable to support the data scale and consistency that AI requires. FHIR's resource model, RESTful API design, standard terminologies, and profile-based governance represent a genuine step-change in what is possible. Regulatory mandates have accelerated adoption past the point where any major U.S. healthcare organization can treat FHIR as optional.

For managers, the practical agenda that flows from this chapter is clear. First, confirm that your core EHR and clinical systems expose FHIR R4 APIs with US Core profile conformance — and verify this independently, not through vendor marketing. Second, identify the FHIR resource types your AI use cases depend on and assess whether your current data sources populate those resources with the fidelity required. Third, build FHIR profile conformance into procurement standards and vendor contracts in the future. Fourth, invest in your team's FHIR literacy — clinicians, data engineers, and product managers all benefit from understanding the resource model and API patterns.

The following checklist captures the management actions warranted at this stage. These are not theoretical recommendations — they are concrete steps that distinguish organizations that are building operational clarity around FHIR from those still treating it as a future-state aspiration.

- Audit your current EHR and data system landscape for FHIR R4 conformance and US Core profile coverage.
- Map your AI use case data requirements to specific FHIR resource types and confirm population fidelity with your IT and data engineering teams.
- Update vendor procurement standards to require FHIR R4 and specified implementation guide conformance as contract terms.
- Evaluate your FHIR Bulk Data export capability for population-scale AI training and analytics pipelines.
- Assess your patient-facing FHIR access posture: app vetting policies, consent management, and patient education materials.
- Identify at least one pilot use case that can leverage FHIR APIs directly — a SMART on FHIR application, a FHIR-based data pipeline, or a FHIR-enabled public health report — and map out the pilot-to-production pathway.
- Invest in FHIR training for data engineers, clinical informaticists, and product managers on your team.

FHIR does not guarantee AI success. It guarantees that the data infrastructure needed for AI success is achievable.

The organizations that move decisively on FHIR governance and architecture now will have a meaningful head start when AI applications reach production scale. The organizations that wait will find themselves rebuilding data pipelines under time pressure — an expensive and avoidable position.

Chapter 3 moves from the FHIR standard itself to the authorization layer that governs who can access FHIR data and under what conditions. SMART on FHIR, OAuth 2.0, and OpenID Connect are the security and identity frameworks that make FHIR APIs usable in production healthcare environments. Understanding them is the next management competency in building a mission-aligned AI and interoperability program.

3 SMART on FHIR: The App Layer That Makes Innovation Possible

Most healthcare organizations have experienced the same exhausting cycle: a promising digital tool emerges, it looks impressive in a demo, and then months of custom integration work follow before a single clinician can use it in a live environment. By the time the app is deployed, the vendor has released a new version, the integration is already brittle, and the project team has moved on to the next emergency. This is not a technology failure — it is an architecture failure. The underlying systems were never designed to support pluggable, interoperable applications.

SMART on FHIR changes that architecture. It is a framework that sits on top of FHIR and defines how applications launch, authenticate, request access, and interact with patient data — regardless of which EHR vendor or health system is running underneath. For healthcare leaders, SMART on FHIR is not an acronym to memorize. It is a procurement philosophy, an operational commitment, and a platform strategy rolled into one specification.

This chapter explains what SMART on FHIR is, how it works technically at a level that informs decisions without requiring a developer background, and why it is the missing piece that connects FHIR data infrastructure to scalable AI deployment. Every AI-powered clinical tool, every decision-support app, every ambient-intelligence workflow you will build or purchase in the next decade will depend on getting this layer right.

3.1 Why SMART on FHIR Exists

3.1.1 The Need for a Universal App Ecosystem

Before SMART on FHIR, the healthcare application market was fragmented by design — not because vendors intended harm, but because no common app layer existed. Each EHR vendor had its own developer program, proprietary APIs, authentication scheme, and data model. A developer who built a clinical risk stratification tool for Epic had to rebuild it almost entirely to work with Cerner, and then again for Meditech. Health systems that wanted to offer their clinicians a best-of-breed portfolio of applications had to maintain a separate integration project for each one.

This fragmentation had a direct cost. A 2019 study published in Health Affairs estimated that US hospitals and physician practices spent more than thirty-one billion dollars annually on health IT administration — much of it consumed by bespoke integration work that produced no clinical value. Every dollar spent wiring one system to another was a dollar not spent on patient care, staff training, or workflow redesign.

The deeper problem was systemic. When every integration is custom, the network effect that makes ecosystems valuable never materializes. App vendors couldn't build a viable business targeting one EHR at a time. Health systems couldn't benefit from the broad app markets that had transformed consumer technology. Innovation was locked behind proprietary walls that even well-intentioned organizations couldn't scale past.

SMART on FHIR was designed to break that lock. The name is an acronym — Substitutable Medical Applications, Reusable Technologies — and it's the mission: substitutable apps, meaning any compliant app can run in any compliant EHR, and technologies that are reusable, meaning developers build once and deploy anywhere. The framework was first developed at Boston Children's Hospital and Harvard Medical School around 2010, incubated within the HL7 standards community, and has since been formally incorporated into ONC's certification requirements for EHRs in the United States. Compliance is no longer optional for major EHR vendors — it is a federal regulatory requirement.

Diagram 3.1 – The Pre-SMART Integration Problem vs. The SMART Solution

Executive Minimalist Infographic Innovation

Before SMART

With SMART on FHIR

EHR vendor silos

EHR vendor silos

Costly custom integration

SMART on FHIR Standard Layer.

3.1.2 Moving from Custom Integrations to Plug-and-Play

The analogy that most healthcare leaders find immediately useful is the smartphone app store. When Apple launched the App Store in 2008, it did not build every application itself. It defined a standard platform — APIs, security requirements, and a distribution mechanism — and then allowed tens of thousands of developers to build against it. The result was an ecosystem that no single company could have produced on its own. SMART on FHIR applies the same principle to healthcare. The EHR becomes the platform. Third-party developers — including AI vendors — build to the SMART standard. Health systems and clinicians access the resulting apps without managing each integration individually.

Plug-and-play is no hyperbole when describing what SMART on FHIR enables in a well-implemented environment. A clinician using an Epic EHR can launch a SMART-enabled AI risk stratification tool from within their chart workflow, have that tool automatically receive the patient's current context — demographics, diagnoses, medications, recent labs — authenticate securely, run its analysis, and return results to the clinician, all without the clinician ever leaving the EHR interface. The entire sequence takes seconds. The integration that makes it possible was built once by the app developer and certified once against the SMART specification, not negotiated separately with every health system that wants to use it.

For managers, the operational clarity this creates is significant. Procurement becomes more predictable — if a vendor's app is SMART-certified and your EHR is SMART-compliant, the integration baseline is known. Contracting can focus on clinical performance and data governance rather than on months of integration scoping. Onboarding timelines compress. And when you want to replace one app with a better one, substitutability means you don't inherit a custom integration debt that makes switching prohibitively expensive.

This is a fundamental shift in how healthcare organizations should think about their application portfolios. The question is no longer 'Can we integrate this?' The question is 'Does this meet the SMART standard?' That shift from capability uncertainty to standards compliance is one of the most practically impactful changes SMART on FHIR delivers.

That said, plug-and-play is a description of the architecture's potential, not a guarantee of a frictionless deployment. Real-world implementations still require clinical configuration, workflow alignment, staff training, and governance decisions about which apps are permitted in which contexts. SMART on FHIR removes the integration barrier. It does not remove the organizational change management work required to determine whether an app actually improves care.

3.2 How SMART on FHIR Works

Understanding how SMART on FHIR works at a technical level — not deeply, but accurately — matters for healthcare managers because it shapes how you evaluate vendors, negotiate contracts, assess security posture, and set expectations with clinical and IT stakeholders. The framework has three core components: launch contexts, scopes, and permissions, and a set of deployment patterns for different user types. Each component has direct management implications.

3.2.1 Launch Contexts

A SMART launch is the sequence of events that occurs when an application is opened within — or in connection with — an EHR session. The concept of a launch context answers a deceptively important question: when the app opens, what does it already know? In a traditional web application, an app knows nothing about the user until it asks them to log in and specify what they want. In a SMART launch, the EHR passes contextual information to the app at launch — typically, which patient is being viewed, which encounter is active, which user is logged in, and which EHR system is the source of truth.

This context transfer is what makes SMART apps feel seamlessly embedded in clinical workflows rather than like separate tools that require re-orientation. A physician viewing a patient's chart clicks a SMART app button. The EHR generates a launch token that encodes the patient ID, the encounter ID, and the physician's user ID. The app receives that token, exchanges it for an access credential, and immediately opens in the context of that specific patient and encounter. The physician never types a patient name, never re-authenticates, and never loses their place in the clinical workflow.

There are two primary launch types. The EHR launch — sometimes called an embedded launch — occurs when the app is launched from within the EHR interface, typically via a button or menu item in the chart. The standalone launch occurs when a user opens the app independently, outside the EHR, and the app then connects to the FHIR server to establish context. Each launch type has different appropriate use cases. EHR launches are ideal for point-of-care clinical tools that need immediate patient context. Standalone launches are appropriate for analytics dashboards, population health tools, care coordination platforms, and patient-facing applications that don't live inside the clinical workflow.

For managers evaluating a SMART-enabled tool, the launch type matters because it determines where the app lives in the clinical workflow and how much friction exists between a clinician's intent and the app's activation. An AI-powered sepsis alert that requires a standalone launch is structurally less likely to be used at the moment of need than one that launches from within the patient chart via an EHR launch. Workflow-level impact starts with the architecture of the launch sequence.

Diagram 3.2 – SMART EHR Launch Sequence

3.2.2 Scopes and Permissions

SMART on FHIR uses a permission model called scopes to control exactly what data an application can access. A scope is a structured string that specifies a resource type, an access level, and a context. For example, a scope of patient/Observation.read means the app is requesting read access to Observation resources — lab results, vital signs — for the currently in-context patient. A scope of the user/Patient. Read means the authenticated user's own patient records. A scope of system/MedicationRequest.Read means backend system access to medication requests, without a user in the loop.

This granularity is not incidental — it is central to responsible deployment. When an AI vendor presents a SMART application for use in your health system, one of the first questions your security team should ask is: what scopes does this application request? An ambient listening tool for surgical documentation has no clinical reason to request read access to mental health or reproductive health records. A population health analytics platform has no reason to request write access to individual patient records. Scope requests that exceed clinical necessity are an immediate red flag, both from a privacy standpoint and as an indicator of an application architecture that was not designed with data minimization in mind.

The SMART specification also supports a consent layer. When a patient-facing application requests scopes on a patient's behalf, the authorization server can present the patient with a clear, plain-language list of what the app wants to access, and the patient can approve or deny individual scopes. This is the same pattern used in consumer app ecosystems — a user installing a fitness app sees a screen that says 'This app wants to access your location and health data' and can choose which permissions to grant. In healthcare, where data sensitivity is higher and the stakes of exposure are real, this consent layer is not cosmetic. It is a substantive protection, and organizations deploying patient-facing SMART apps should verify that their authorization server implementation presents consent screens that are genuinely comprehensible to patients — not buried in technical jargon.

From a governance perspective, scope management is one of the most actionable tools healthcare organizations have for controlling third-party application access to clinical data. Many FHIR server implementations allow administrators to pre-configure which scopes a given application is permitted to request, creating an organizational-level policy layer on top of the application's own scope requests. This is responsible by design — building the access control into the infrastructure rather than relying solely on the application vendor's self-reported claims about what their software does with data.

Scope management also has a contractual dimension. When negotiating with AI application vendors, organizations should require that the application's scope requirements be documented in the contract, that any scope changes require organizational approval, and that the vendor's data use agreement explicitly limits their use of received data to the stated clinical purpose. Scopes are not just a technical control — they are the technical expression of a data governance agreement.

3.2.3 Patient-Facing, Provider-Facing, and Backend Services

SMART on FHIR supports three distinct deployment patterns, each designed for a different user type and use case. Understanding these patterns helps managers match the right architecture to the right clinical problem and avoid deploying a tool in a context for which it was not designed.

Patient-facing SMART apps are designed to run within a patient's access to their health data, typically via a patient portal or a mobile health application. These apps authenticate the patient as the user, request scopes on the patient's behalf, and operate on data the patient has the right to access under HIPAA and the information-blocking rules of the 21st Century Cures Act. Examples include personal health record apps, medication management tools, chronic disease monitoring platforms, and mental health support applications. Patient-facing apps are often the most visible category to the public and carry the highest reputational risk if they mishandle data.

Provider-facing SMART apps are designed to run within the clinical workflow and are authenticated as the treating clinician or care team member. They are the most common category in hospital and clinic deployments and include clinical decision support tools, order management assistants, care gap alerts, and AI-powered risk scoring applications. Provider-facing apps operate with the permissions of the authenticated user — a physician's app access is scoped to what that physician is authorized to see in the EHR — creating a natural alignment between SMART access controls and existing role-based access policies.

Backend service SMART apps — sometimes called system-to-system or headless apps — operate without a human user in the active session. They authenticate directly with the FHIR server using machine credentials, typically via the SMART Backend Services authorization profile. Population health analytics engines, automated quality reporting systems, AI model training pipelines, and overnight batch processing jobs are common backend service use cases. These apps are powerful and often invisible to clinical staff, which makes their governance particularly important. An organization may have patient-facing and provider-facing apps that are carefully reviewed and approved, while a backend service quietly exports millions of records for an analytics partner. Backend SMART apps require the same governance rigor as any other category — and in some respects, more, because the absence of a human user in the session means that human oversight cannot substitute for technical controls.

Diagram 3.3 – Three SMART Deployment Patterns

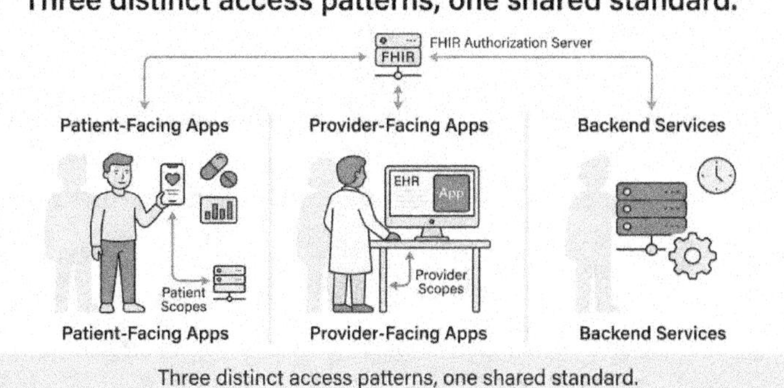

Three distinct access patterns, one shared standard.

3.3 SMART as an Innovation Platform

A standard that only solves an integration problem is useful. A standard that creates a platform for sustained innovation is transformative. SMART on FHIR has shifted from the former to the latter over the past decade, and understanding what that shift means for healthcare organizations — practically, not theoretically — is essential for leaders building their digital health strategy.

3.3.1 How Developers Build Once and Deploy Everywhere

The 'build once, deploy everywhere' principle is the economic engine behind the SMART ecosystem. When a developer builds a SMART-compliant application, they are investing in a single codebase that can, in principle, be deployed to any FHIR-enabled EHR. That fundamentally changes the economics of healthcare app development. A developer building a sepsis risk prediction tool no longer needs to budget for separate integrations with Epic, Oracle Health, Meditech, and MEDITECH Expanse. The SMART layer handles the authentication, the context passing, and the data retrieval interface. The developer focuses their resources on the clinical logic and the user experience.

For health systems, this economic shift translates into a broader and more competitive application market. When development costs are lower, more developers can afford to enter the market, including smaller companies and academic innovators who would never survive the integration cost of the old model. More competitors mean more options, better quality, and more price pressure. Organizations that have shifted their procurement strategy to favor SMART-compliant tools consistently report shorter vendor selection cycles, more competitive pricing, and a larger pool of qualified vendors to evaluate.

The caveat that matters for managers is the word 'principle.' SMART compliance means the authentication and authorization layer is standardized. It does not mean the application will work identically across every EHR with zero configuration. EHR vendors implement FHIR differently. Some expose a richer set of FHIR resources than others. Some support the latest SMART specification versions, while others lag. A SMART app built for a fully FHIR R4-compliant environment may need to be adjusted to run on an EHR that exposes only a subset of R4 resources. Due diligence still requires asking vendors which FHIR resources their apps depend on and verifying that your EHR exposes those resources at the required level of fidelity.

That said, the gap between 'in principle' and 'in practice' has narrowed substantially over the past five years as EHR vendors have invested in their FHIR implementations under regulatory pressure. Organizations that are evaluating SMART-based tools today face far fewer interoperability surprises than those who explored the market in 2018. The pilot-to-production pathway is more reliable, and the organizational investment required to navigate it is lower.

3.3.2 How SMART Accelerates AI Adoption

AI in healthcare faces a distinct deployment problem distinct from its development problem. Researchers and data scientists can build impressive models in research environments using curated datasets. The hard part is getting those models into clinical workflows where they operate on real-time data, update as patient conditions change, present results to the right clinician at the right moment, and integrate into the documentation and ordering systems that clinicians actually use. That last mile — from a trained model to a workflow-level impact — is where most healthcare AI efforts stall.

SMART on FHIR directly addresses the last-mile problem. An AI model deployed as a SMART application can receive real-time patient context at launch. It can query the FHIR server for precisely the data it needs using standardized FHIR queries. It can return results in a structured format that the EHR can display within the chart, include in a CDS Hooks response, or route to a specific care team member. The entire pipeline from data to inference to display operates within the SMART framework, without requiring a custom integration for each step.

This is not a minor convenience. It changes the timeline for deploying AI from a multi-year integration project to weeks or months for organizations with a well-configured FHIR infrastructure. It changes the risk profile because a SMART-based deployment can be scoped, permissioned, and monitored using the same tools the organization already uses for other SMART applications. And it changes the accountability structure, because the scopes the AI app requests are explicit and auditable — a manager can look at a SMART app's scope list and have a meaningful conversation about what data that AI system can and cannot access.

Perhaps most importantly for mission-aligned AI deployment, SMART on FHIR enables organizations to integrate AI tools from multiple vendors without each one becoming its own bespoke integration project. A health system can run a SMART-based AI for sepsis prediction from one vendor alongside a SMART-based AI for medication reconciliation from another, both running within the same EHR environment, both governed by the same authorization server and scope policies, both auditable through the same access logs. Composability — the ability to assemble a portfolio of AI capabilities from multiple sources — is one of SMART on FHIR's most strategically valuable properties.

3.3.3 Examples of SMART-Enabled Apps

The SMART App Gallery, maintained by the SMART Health IT project, lists hundreds of applications across clinical, administrative, and patient-facing categories. The range is instructive — it demonstrates both the maturity of the ecosystem and the breadth of clinical problems to which it is being applied.

In the clinical decision support category, SMART-enabled tools include sepsis early warning systems, acute kidney injury alerts, antimicrobial stewardship advisors, and drug-drug interaction checkers. These tools pull real-time patient data from the FHIR server, apply a clinical algorithm or AI model, and surface results within the EHR as alerts, recommendations, or visual dashboards — without requiring the clinician to navigate to a separate system.

In the population health category, SMART backend services are used for care gap identification, chronic disease registries, aggregation of social determinants of health screening, and value-based care performance tracking. These tools typically run as scheduled backend processes, querying the FHIR server for cohort-level data and returning structured outputs to population health management platforms or EHR-embedded dashboards.

In the patient-facing category, personal health record aggregators, medication adherence tools, remote monitoring platforms for conditions like heart failure and diabetes, and post-discharge care coordination apps all use the SMART patient-facing launch pattern. Under the ONC's information-blocking rules and the CMS patient access API requirements, health systems are now obligated to support patient access through SMART-compliant third-party apps — making this category not just an innovation opportunity but a compliance requirement.

The operational lesson for managers is that the SMART ecosystem is not speculative. Mature, clinically validated applications exist across every major care domain. An organization that has invested in well-configured FHIR and

SMART infrastructure is not waiting for the market to develop — it is plugging into a robust, rapidly growing market.

Diagram 3.4 – The SMART App Ecosystem by Category

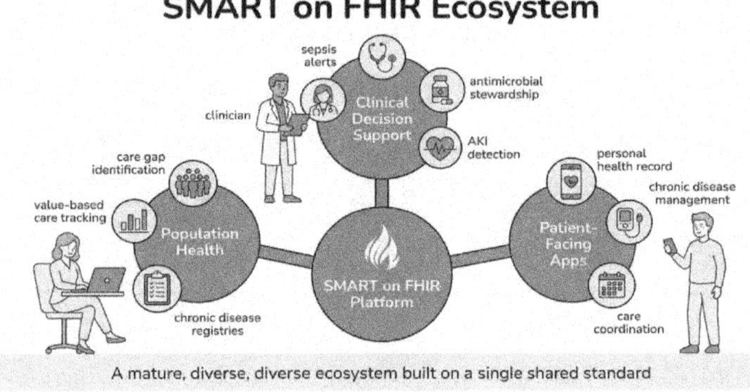

3.4 SMART + AI

The convergence of SMART on FHIR and artificial intelligence is not a future scenario — it is happening now in organizations across the country. What varies is whether that convergence is happening with deliberate governance or by default. The organizations that will get the most value from AI in clinical settings are those that have intentionally built their SMART infrastructure with AI deployment in mind and that apply the same disciplined decision-making to AI app procurement that they would to any other clinical tool.

This section examines the four primary categories where SMART-enabled AI is producing workflow-level impact today, and the management considerations that determine whether each category delivers value or creates risk.

3.4.1 AI-Powered Clinical Decision Support

Clinical decision support (CDS) has existed in EHRs for decades in the form of rule-based alerts — if a patient's potassium is above 5.5, fire an alert. The problem with rule-based CDS is well-documented: alert fatigue. When every rule fires independently, clinicians receive hundreds of alerts per shift, the vast majority of which are either already known, clinically irrelevant to the patient's current situation, or so low-risk that they don't change the plan of care. Studies have consistently found that clinicians override 70 to 90 percent of EHR alerts. A system that generates thousands of alerts per day but changes clinical behavior in fewer than ten percent of cases is not delivering decision support — it is delivering noise.

AI-powered CDS breaks from this pattern by using machine-learning models trained on large clinical datasets to generate risk scores and recommendations that are contextually relevant to the specific patient at the specific moment of care. Instead of firing every time potassium exceeds a threshold, an AI-powered CDS tool might integrate the potassium level with the patient's creatinine trend, current medication list, cardiac history, and fluid balance to produce a composite risk score that fires only when a clinically meaningful intervention is warranted.

SMART on FHIR is what makes this possible at scale. The AI model, deployed as a SMART app or integrated via the CDS Hooks standard — which itself builds on the

SMART/FHIR stack — receives real-time patient data as soon as a clinical event occurs. A physician opens a medication order. The EHR calls a CDS Hooks endpoint. The AI model receives the patient's current medication list and relevant clinical data via FHIR, evaluates the order against its trained model, and returns a structured recommendation within milliseconds. The physician sees a targeted, clinically relevant alert rather than a boilerplate drug interaction warning.

For managers, the performance metrics to track for AI-powered CDS differ from those for traditional CDS. The key indicators are alert specificity (what percentage of alerts fired led to a clinical action), override rate (what percentage of alerts are dismissed without action — the lower, the better), and outcome correlation (does firing the alert actually correlate with better patient outcomes in your population). These metrics require a data infrastructure capable of linking alert events to downstream clinical actions and patient outcomes — another area where a well-configured FHIR server provides foundational support.

Governance questions for AI-powered CDS also need to be on the manager's agenda from the start. Who is accountable when an AI alert fires and a clinician follows its recommendation — and the outcome is poor? What is the process for updating the model when clinical evidence evolves? How is the model's performance monitored for drift — the gradual degradation in accuracy that occurs as the patient population or care patterns change? These are not hypothetical concerns. There are active regulatory and liability questions that the FDA, ONC, and CMS are actively developing guidance on.

3.4.2 Ambient Intelligence

Ambient intelligence represents one of the most practically impactful applications of AI in healthcare, and also one of the least understood by most healthcare managers. The core concept is simple: AI that operates in the background of clinical environments, continuously sensing and processing information without requiring explicit clinician interaction, and surfacing insights when they are relevant.

The most mature commercial application of ambient intelligence in healthcare today is ambient clinical documentation — AI systems that listen to the physician-patient encounter, transcribe and structure the conversation, and automatically draft clinical notes in the EHR. These systems, offered by vendors including Nuance (Microsoft), Suki, Abridge, and others, have demonstrated significant reductions in documentation time and improvements in physician satisfaction with EHR usability. Studies have shown reductions in documentation time of 40 to 60 percent, with physicians reclaiming significant time previously lost to keyboard work after patient encounters.

The SMART connection is direct. Ambient documentation AI systems integrate with EHRs via SMART on FHIR to write completed note drafts back to the patient's chart, pre-populate structured data fields like diagnoses and medications, and trigger downstream documentation workflows. The note draft created by the ambient AI is written to a FHIR DocumentReference or Composition resource, which the EHR retrieves and presents to the physician for review and signature. Without the SMART/FHIR layer, this write-back would require a custom integration for each EHR vendor — making deployment slow, expensive, and difficult to scale.

Beyond documentation, ambient intelligence applications in development or early deployment include continuous patient monitoring systems that detect deterioration signals from physiological sensors without requiring manual nursing assessment triggers, ambient room awareness systems in ICUs that monitor patient movement and alert staff to fall risk or line dislodgement, and ambient OR intelligence that tracks surgical workflow stages and surfaces relevant imaging or laboratory data proactively.

The management challenge with ambient intelligence is consent, transparency, and policy clarity. Recording conversations in clinical settings requires patient consent in most jurisdictions and raises legitimate concerns about the privacy of sensitive disclosures. Organizations deploying ambient documentation AI must have a clear consent workflow, a policy for what happens when patients decline, a documented process for how recorded audio is stored and deleted, and a contractual commitment from the vendor that audio data is used only for documentation, not for model training, without explicit consent. These are not optional considerations — they are requirements for responsible deployment.

3.4.3 Predictive Workflows

Predictive workflows use AI models to anticipate clinical events or operational needs before they occur, enabling proactive rather than reactive care delivery. The category is broad — it includes clinical prediction models for readmission risk, sepsis onset, deterioration, and acute kidney injury, as well as operational models for patient flow, bed management, staffing demand, and supply chain optimization.

The clinical prediction use case is most directly connected to SMART on FHIR. A readmission risk prediction model, for example, needs access to a patient's diagnosis codes, medication list, prior visit history, social history, and recent laboratory results at or near the time of discharge. All of those data elements are available as FHIR resources. A SMART backend service can query for them on a scheduled basis — say, every hour for all patients flagged as high-risk — run the prediction model, and write the resulting risk score back to the EHR as a FHIR Observation or Flag resource, where it triggers a care coordinator workflow.

The operational prediction use case, while less direct in its reliance on FHIR, benefits from the same underlying data infrastructure. A patient flow optimization model needs census data, admission and discharge timestamps, procedure scheduling data, and emergency department arrival patterns — all of which can be exposed as FHIR resources and queried by an analytics engine. Organizations that have invested in a comprehensive FHIR data layer find that their operational AI initiatives benefit from the same infrastructure they built for clinical AI, creating compounding returns on the initial investment.

The critical governance question for predictive workflows is what happens when the prediction fires. A sepsis prediction score that appears in the EHR without a clear recommended action, a designated responder, or a feedback mechanism to capture whether the action was taken and what the outcome was is a prediction without a workflow. It generates data without generating value. Organizations that have successfully operationalized predictive AI are those that designed the entire workflow — trigger, alert, recommended action, responder, response documentation, outcome tracking — before go-live, not after.

SMART on FHIR supports this end-to-end design. The FHIR server can be the source of the input data, the repository for the prediction output, the trigger for the alert workflow, and the repository for the documented response. Building the workflow within the FHIR data layer, rather than in a parallel system, means the entire chain of events is auditable, reportable, and available for retrospective analysis of model performance.

Diagram 3.5 – AI-Powered Predictive Workflow via SMART on FHIR

Closed-loop, Auditable, FHIR-native Predictive Workflow

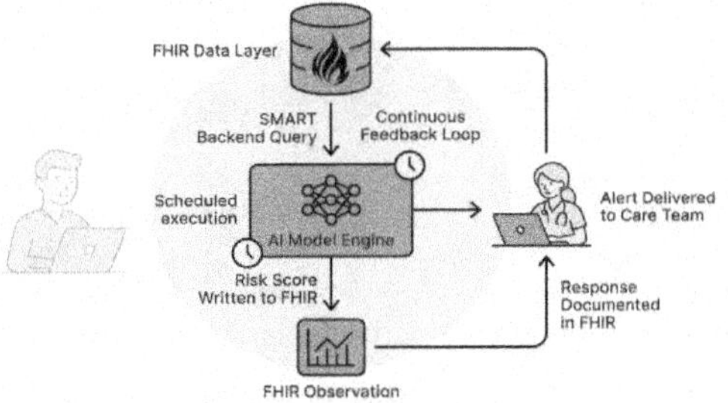

Executive infographic - Closed-loop, auditable, FHIR-native predictive workflow

3.4.4 Why SMART Is Essential for Scalable AI Deployment

The question of scalability is ultimately whether an organization can add, modify, and retire AI capabilities without incurring prohibitive overhead for each change. In the pre-SMART world, each AI application was a custom integration — adding a new one meant months of work, retiring an underperforming one meant untangling dependencies, and updating a model meant coordinating changes across multiple proprietary interfaces. The marginal cost of each AI deployment was high enough to limit most organizations to one or two AI pilots at a time.

SMART on FHIR changes the scalability equation. When AI applications are built to the SMART standard and the EHR is a compliant SMART platform, the organization can manage a portfolio of AI applications the same way it manages other application portfolios — through a centralized configuration, a defined approval process, a shared authorization server, and standardized audit logging. Adding a new AI app does not require a new integration project. It requires an organizational review and approval of the app's scope request, a clinical configuration of its launch context, and a communication workflow for the relevant care team. These are manageable, repeatable activities.

This scalability is mission-aligned in a specific, practical sense. Healthcare organizations face AI adoption pressure from multiple directions simultaneously — from payers who want AI-enabled quality reporting, from clinical staff who want AI-assisted documentation, from patients who want AI-powered navigation tools, and from administrators who want AI-driven operational efficiency. An organization that can manage only one or two AI deployments at a time will struggle to respond effectively to all of these pressures. An organization with a mature SMART infrastructure can add AI capabilities incrementally, in a governed and auditable way, without each new deployment consuming the full attention of its IT and clinical leadership.

There is also a vendor independence dimension to scalability that managers should weigh carefully. An organization that deploys AI exclusively through proprietary EHR-native tools is dependent on the EHR vendor's AI development roadmap, pricing, and capabilities. A SMART-based strategy allows the organization to select best-in-class AI capabilities from a diverse vendor pool, replacing underperformers without disrupting the underlying infrastructure. This is a form of strategic risk management — distributing AI capability sourcing across multiple vendors, in the same way a prudent organization diversifies its critical supplier relationships.

The organizations that are most advanced in AI deployment today — major academic medical centers, integrated delivery networks with strong informatics leadership, and forward-thinking community health systems — are consistently those that invested in SMART on FHIR infrastructure before they began evaluating specific AI applications. The infrastructure investment preceded the application portfolio because the leaders who made those decisions understood that a sound platform is what makes an app ecosystem possible, not the other way around.

Diagram 3.6 – SMART as the AI Deployment Platform

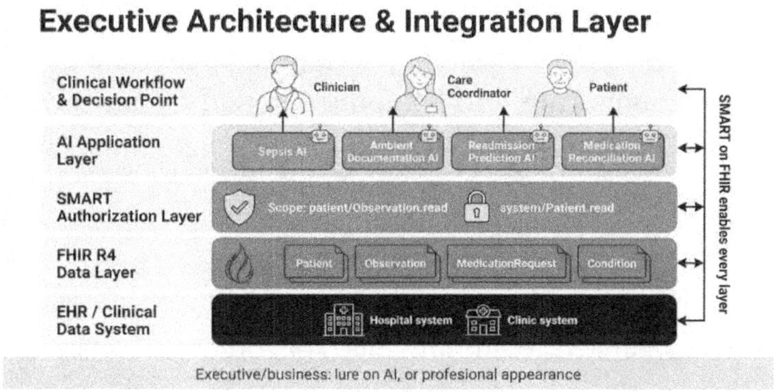

3.5 Takeaway

SMART on FHIR is the app layer that transforms FHIR from a data standard into a clinical innovation platform. It solves the integration problem that has made healthcare application deployment slow, expensive, and fragile. It provides the authentication, authorization, and context-passing infrastructure that AI applications need to operate within clinical workflows in real time. And it does so in a way that is governable, auditable, and scalable — the three properties that separate a sustainable AI strategy from a series of isolated pilots.

For healthcare managers, the SMART framework highlights decisions that belong at the leadership level, not the technical level. The decision to require SMART compliance in all future application procurement is a strategic choice that compresses integration timelines, expands the vendor pool, and creates the conditions for a composable AI portfolio. The decision to configure scope policies on your FHIR authorization server is a governance choice that defines what data any third-party application — including AI systems — can access and under what conditions. The decision to build outcome tracking and model monitoring into AI application governance from the start is a quality-and-safety choice that determines whether your AI investments deliver sustained clinical value or gradually drift into producing noise.

The following checklist captures the highest-priority SMART-related commitments for organizations building toward a scalable AI capability:

- Confirm that your EHR vendor's FHIR R4 implementation supports SMART on FHIR 2.0 (the current specification version), including both EHR launch and standalone launch patterns.
- Implement a centralized SMART authorization server with organizational scope policies — do not allow apps to request arbitrary scopes without administrative review and approval.

- Add SMART compliance as a contractual requirement in all future application and AI vendor agreements, alongside documented scope requirements and data use restrictions.
- Establish a SMART app governance process that mirrors your medical device or clinical tool approval process — clinical review, security review, privacy review, and a defined monitoring cadence post-deployment.
- Design the full clinical workflow — trigger, alert, recommended action, responder, response documentation, outcome tracking — before deploying any AI application, not after go-live.
- Audit your existing application portfolio for SMART compliance gaps, particularly any backend service or analytics application that accesses patient data without a human user in the active session.
- For patient-facing SMART apps, verify that your authorization server presents consent screens that are comprehensible to patients and accurately describe what data the app will access.

SMART on FHIR did not receive the same public attention as the EHR incentive programs or the information-blocking rules. It has been built into federal certification requirements quietly, through a technical standards process that most healthcare leaders have not followed closely. But its practical impact on what is possible in clinical AI deployment is greater than almost any other interoperability development of the past decade. Organizations that understand it and act on it now will be positioned to move at a speed and scale that those who ignore it cannot match.

The next chapter moves from the application layer to the intelligence layer — examining how AI models are trained, validated, and deployed responsibly within the FHIR and SMART infrastructure described in this chapter. The architecture is in place. The question becomes: what AI goes in it, and how do you know it is safe?

4 OAuth and OpenID Connect: The Trust Framework

Every healthcare AI deployment ultimately reduces to a question of trust. Clinicians must trust that an AI recommendation is grounded in complete, accurate patient data. Patients must trust that their most sensitive information is not accessible to unauthorized systems. Regulators must trust that access controls are auditable, enforceable, and aligned with the law. And IT leaders must trust that the security architecture they put in place today will hold as systems scale, integrations multiply, and AI becomes more deeply embedded in care delivery.

OAuth 2.0 and OpenID Connect are the industry-standard protocols that enable this trust. They are not academic abstractions — they are the working infrastructure that controls who can access which data, under what conditions, for how long, and on whose behalf. If FHIR is the language that healthcare data speaks, OAuth and OpenID Connect are the locks and keys that determine who gets to read that language and when.

This chapter explains both protocols in plain language, connects them directly to the realities of clinical workflow, and frames the decisions healthcare leaders need to make when deploying AI systems that touch protected health information. The goal is not to make you an OAuth engineer. It is to give you enough operational clarity to evaluate vendors, challenge security assumptions, and make responsible-by-design choices about how AI interacts with your data environment.

4.1 Why Security Matters in Healthcare

4.1.1 Sensitivity of Clinical Data

Healthcare data is among the most sensitive information a person generates in their lifetime. A diagnosis, a medication list, a psychiatric history, a genetic profile — these are not transactional records. They carry real-world consequences for employment, insurance, relationships, and personal safety. A breach of financial data can be remediated. A breach of clinical data cannot be undone. Once a person's HIV status, addiction history, or mental health diagnosis is exposed, the harm is permanent.

This is not a theoretical concern. Healthcare organizations are among the most targeted institutions for cyberattacks. According to breach data published by the U.S. Department of Health and Human Services, the healthcare sector reports more large-scale data breaches than any other industry sector covered under federal breach notification requirements. The average cost of a healthcare data breach consistently ranks highest among industries. And those figures do not capture the harder-to-quantify harms: loss of patient trust, chilling effects on care-seeking behavior, and reputational damage that follows a public disclosure of a breach.

For AI systems, the sensitivity of clinical data creates a specific and compounding risk. AI models often require access to large volumes of historical data for training, validation, and inference. They operate across patient populations, not just individual records. A poorly controlled AI pipeline is not a single-record vulnerability — it is a population-scale exposure risk. Managers must understand that AI amplifies both the value and the risk surface of clinical data, which is precisely why security controls at the access layer are not optional overhead but mission-aligned infrastructure.

Diagram 4.1 – The Healthcare Data Sensitivity Spectrum

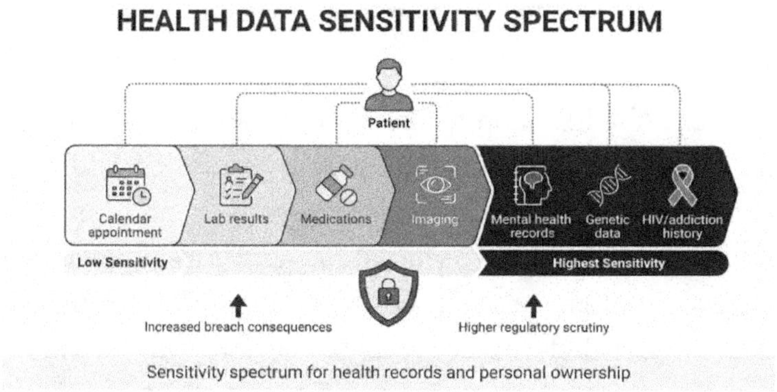

Sensitivity spectrum for health records and personal ownership

4.1.2 Regulatory Expectations

HIPAA's Security Rule requires covered entities and their business associates to implement technical safeguards that control access to electronic protected health information (ePHI). This includes unique user identification, automatic logoff, encryption, and audit controls. These requirements do not specify OAuth or OpenID Connect by name — the regulations are intentionally technology-neutral — but the requirements map directly onto what these protocols provide.

The ONC's 21st Century Cures Act Final Rule and the CMS Interoperability and Patient Access Final Rule added a new dimension: they require that covered APIs use standardized FHIR endpoints and that access to those endpoints be controlled through OAuth 2.0 with SMART on FHIR scopes. This is not guidance — it is regulation. Payers and providers subject to these rules must implement compliant authorization frameworks or face enforcement consequences. The practical result is that OAuth is no longer optional for organizations deploying FHIR-based data exchange.

The GDPR, applicable in European jurisdictions and increasingly influential on global practice, adds further expectations regarding lawful basis for processing, data minimization, purpose limitation, and individuals' right to know how their data is used. AI systems processing clinical data in jurisdictions covered by GDPR must demonstrate that access was lawful, purposeful, and constrained to what was necessary. OAuth's scope system is a direct mechanism for operationalizing these requirements — scopes define purpose, and scope constraints enforce data minimization.

For healthcare managers, the regulatory picture creates both a compliance floor and a strategic framing. The floor is clear: you must implement access controls that meet HIPAA, ONC, and applicable state requirements. The strategic opportunity is that well-implemented OAuth and OIDC go beyond compliance — they create an auditable record of who accessed what, when, and for what purpose, which is exactly what you need to demonstrate accountability to regulators, auditors, and patients.

4.1.3 Trust as a Prerequisite for AI Adoption

In the clinical environment, trust is not a marketing concept. It is the prerequisite for adoption. A radiologist will not rely on an AI diagnostic aid unless they believe the AI reviewed the right patient's complete imaging history. A care manager will not act on an AI-generated risk score unless they trust that the model accessed the full care record, not a subset that excluded recent ED visits or specialist notes. If clinicians suspect that the data pipeline feeding an AI system is incomplete, insecure, or improperly scoped, they disengage — and the investment in that AI system produces no clinical value.

Trust is also a prerequisite for patient willingness to share data. When patients are asked to authorize third-party applications to access their health records — as increasingly happens through patient-facing FHIR APIs — they make a trust decision. They evaluate whether the requesting application is legitimate, whether the scope of access is reasonable, and whether they can later revoke access. OAuth's authorization flows are the visible face of that trust decision. A well-designed OAuth consent experience reinforces trust. A poorly designed one — or one that obscures what is being requested — undermines it.

For AI systems specifically, trust has an additional dimension: explainability of data access. When an AI recommendation is questioned clinically, the care team must be able to trace not just the logic of the recommendation but the completeness and legitimacy of the data that informed it. OAuth access logs and token audit trails become part of that explanation. The pilot-to-production pathway for clinical AI systems should include a defined plan for how data access is logged, reported, and available for review when recommendations are challenged.

4.2 OAuth Explained Simply

4.2.1 Authorization vs. Authentication

OAuth 2.0 is an authorization protocol. It answers the question: What is this application allowed to do? It does not, by itself, answer: who is this person? That distinction matters enormously in practice, and confusing the two leads to security gaps.

Authentication is the process of verifying identity — confirming that the person or system claiming to be User A is actually User A. Authorization is the process of determining what that verified identity is permitted to do — read lab results, submit orders, access administrative data. You can have authorization without strong authentication, which is a vulnerability. You can have authentication without scoped authorization, which means an authenticated user gets access to everything, which is also a vulnerability.

In the healthcare context, this distinction has a direct workflow-level impact. A patient portal user authenticates with their username and password. OAuth then controls what that authenticated user can authorize a third-party app to do with their data. The app does not receive the patient's credentials — it receives a scoped access token that represents the patient's permission for a specific action. This is the fundamental design insight of OAuth: it allows delegation of access without delegation of credentials.

For AI systems operating in the background — without a human user actively approving each action — the pattern is slightly different. The AI system authenticates as a trusted application using credentials registered with the authorization server, then requests the scoped access it needs to the data it requires. The authorization logic still applies: the system can only access data it has been explicitly permitted to access, regardless of how much data technically exists in the FHIR repository. This is the core control that makes AI deployment responsible by design.

Diagram 4.2 – Authentication vs. Authorization: The Two-Gate Model

4.2.2 Access Tokens, Refresh Tokens, and Scopes

When an OAuth flow completes successfully, the application receives an access token. Think of this as a time-limited parking pass for a specific lot. It grants access to defined resources for a defined period. When the parking pass expires, the applicant cannot park anymore — they must get a new one. The access token does not contain the user's credentials. It contains encoded information about what is permitted and for how long.

Access tokens are intentionally short-lived. This is a security feature, not an inconvenience. If an attacker intercepts an access token, the window of damage is limited to the token's remaining lifespan — often fifteen minutes to an hour. After expiration, the token is useless. For AI systems that need continuous or long-running access to data, the refresh token is the mechanism for obtaining new access tokens without requiring the user to re-authenticate. The refresh token is longer-lived but is held securely by the application, not transmitted with every request.

Scopes are the granular specification of what an access token permits. In the SMART on FHIR framework, scopes follow a defined syntax: patient/Observation.read means the token permits reading Observation resources for the patient who authorized access. user/MedicationRequest.write means the token permits a credentialed user to write medication request resources. system/Patient. read means the token permits a backend system to read Patient resources without a human user in the loop.

The scope system is where data minimization becomes operational. An AI system that needs to read laboratory results to generate a risk score should request only the scopes necessary for that function — not administrative write access, not full patient demographics, not medication orders unless those are directly required. Healthcare managers overseeing AI vendor deployments should review the scope list requested by every AI system and ask specifically: why does this system need this scope? If the vendor cannot explain the necessity of each requested scope, that is a red flag.

Practical scope governance is not a one-time task. As AI systems evolve and their functions expand, the scope of requests expands too. Without active oversight, scope creep leads to systems that access far more data than they need, violating data minimization principles, increasing breach exposure, and potentially introducing regulatory violations. The pilot-to-production pathway for any AI system should include a defined process for scope review at each stage of deployment, not just at initial registration.

4.2.3 How OAuth Protects Data Access

OAuth's protective value rests on four interconnected properties: delegation without credential sharing, time-limited access, scope constraints, and revocability. Each of these properties addresses a specific class of risk.

Delegation without credential sharing means that when a patient authorizes a nutrition app to read their lab results from the hospital's FHIR API, the nutrition app never receives the patient's hospital login credentials. It receives a scoped token. If the nutrition app is later compromised, the attacker cannot use the patient's hospital credentials to access other systems — because those credentials were never shared. The damage is contained to what the token permitted.

Time-limited access means that tokens expire. A token issued to support a thirty-day AI analysis project does not persist indefinitely. Expiration is enforced automatically, without requiring anyone to remember to revoke access. For healthcare managers, this is operationally significant: it means that short-term AI projects or vendor pilots do not leave persistent access grants in your environment that outlast the project itself — provided the token lifetimes are configured correctly.

Scope constraints mean that access is bounded by what was explicitly requested and approved. A system granted patient/Observation.read cannot use that token to write records or access resources outside the approved scope, regardless of what the underlying FHIR server technically contains. Scopes are enforced at the authorization server level, not solely at the application level — meaning even a compromised application cannot exceed its authorized scope.

Revocability means that access grants can be canceled at any time by the authorizing party — the patient, the administrator, or the system owner. When a vendor contract ends, when a clinical trial concludes, or when a security incident is detected, access can be revoked by invalidating the tokens and refresh tokens associated with that application. This is a critical operational control. Managers should confirm with their identity and access management teams that revocation is not just theoretically possible but is operationally rehearsed and can be executed within minutes when needed.

Diagram 4.3 – The OAuth Authorization Flow for a Clinical AI Application

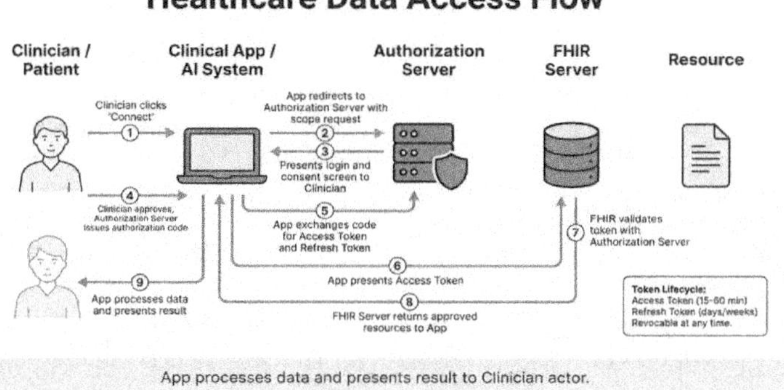

4.3 OpenID Connect Explained Simply

4.3.1 Identity Verification

OpenID Connect (OIDC) is a thin identity layer built on top of OAuth 2.0. Where OAuth answers 'what is this application allowed to do,' OpenID Connect answers 'who is this person, and can we verify it?' OIDC extends the OAuth flow to return not just an access token but also an ID token — a signed, verifiable assertion about the authenticated user's identity.

The practical relationship is this: OAuth 2.0 handles the authorization. OpenID Connect handles the authentication. They run together in the same flow, using the same protocol mechanics, but serve distinct functions. For most healthcare applications — including AI systems that operate on behalf of specific clinicians or patients — you need both: you need to know who is making the request (OIDC) and you need to control what that identified person can do (OAuth).

Identity verification in healthcare is not just a technical checkpoint. It is a clinical and legal requirement. Medication orders, clinical documentation, and diagnostic sign-offs require that the author is identifiable, credentialed, and accountable. An AI system that assists with clinical documentation must be configured to document not just the AI's contribution but the identity of the clinician who reviewed and authorized the output. OIDC provides the identity assertion that makes this attribution traceable.

For AI systems, identity verification extends to the system's own identity. A well-governed AI deployment registers the AI application as a known entity in the organization's identity provider. When the AI system requests access to FHIR data, it presents its own credentials — a client ID and client secret or a signed assertion — that the authorization server validates. This establishes that the request came from the known, registered AI system, not from an unauthorized process or a compromised intermediary. This is the foundation of machine-to-machine trust in clinical environments.

4.3.2 ID Tokens and User Info

The ID token is a JSON Web Token (JWT) — a compact, cryptographically signed data structure that contains claims about the authenticated user. Standard claims include the user's unique identifier (sub), the issuer (iss), the audience the token is intended for (aud), and the token's expiration time (exp). In healthcare deployments, extended claims can include the user's role, their organizational affiliation, their NPI number, or any other attribute the identity provider is configured to assert.

The identity provider's private key signs the ID token. Any system with access to the corresponding public key can verify that a trusted source issued the token and has not been tampered with. This verification is cryptographic — it does not require a call back to the identity provider for every request. The token carries its own proof of authenticity. For high-volume AI applications processing thousands of requests, this is an important performance property: identity can be verified locally without round-trip latency to a central authentication server.

The UserInfo endpoint is a complementary mechanism: an OIDC-protected API that returns additional claims about the authenticated user that were not included in the ID token. An AI application can query the UserInfo endpoint to obtain additional context about the user — their current department, specialty, and active role assignment — that may be relevant to how the AI presents information or what data it prioritizes. This creates a pathway for clinician-context-aware AI behavior that is grounded in verified identity attributes, not self-asserted user preferences.

For healthcare managers, the key governance question around ID tokens is: what claims does your identity provider currently issue, and are those claims sufficient to support the AI workflows you are deploying? If your identity provider does not issue role or specialty attributes, AI systems that need to tailor behavior by clinical role will need a workaround — and workarounds introduce complexity and risk. Aligning your identity provider configuration with your AI use cases is a prerequisite task that should be completed early in any AI deployment plan.

4.3.3 Why Identity Matters for Clinical Workflows

Clinical workflows are identity-dependent in ways that most enterprise software environments are not. In healthcare, the same task — reviewing a medication list — carries different implications depending on whether the reviewer is an attending physician, a pharmacist, a nurse, a medical student, or an AI system. The appropriate response, the level of detail displayed, the actions available, and the documentation requirements all vary by role. A security architecture that cannot distinguish between these identities cannot support safe, compliant clinical AI deployment.

Role-based access control (RBAC) and attribute-based access control (ABAC) systems depend on identity as their input. OpenID Connect provides that input. When a clinician launches an AI-assisted documentation tool, the OIDC ID token identifies the clinician, their role, and the permissions associated with that role. The AI system can then present contextually appropriate information — a physician sees full diagnostic detail, a medical student sees an educational overlay, a patient sees a plain-language summary — all from the same underlying FHIR data, differentiated by verified identity.

Identity also matters for audit. Every clinical system is subject to audit log requirements — who accessed what, when, and for what stated purpose. AI systems that process clinical data must be included in that audit trail. OIDC identity assertions provide the user-level attribution that links an AI-mediated data access event to a specific authenticated user or system. Without this link, audit logs are incomplete, compliance demonstrations are difficult, and incident investigations are guesswork.

Diagram 4.4 – OpenID Connect Identity Flow in a Clinical AI Context

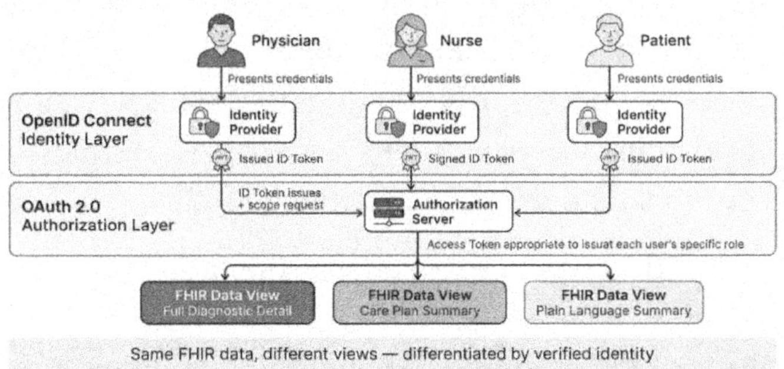

4.4 Trust for AI Systems

4.4.1 Ensuring AI Only Accesses Appropriate Data

The central governance challenge with AI in healthcare is not capability — it is constraint. AI systems can process enormous volumes of data, draw inferences across populations, and operate continuously without human review. That capability is valuable. It is also dangerous if not precisely bounded. The question is not 'can this AI access this data?' but 'should this AI access this data for this purpose?' OAuth's scope framework is the primary technical mechanism for answering that question consistently, auditably, and enforceably.

Appropriate data access for AI systems requires defining three boundaries: the patient boundary, the resource boundary, and the purpose boundary. The patient boundary determines which patients' data the AI can access — is it a population-level analytics system that may access all patients, or a point-of-care tool that should only access the patient currently being treated? The resource boundary determines which FHIR resource types the AI can read or write — a risk stratification model may need Observation, Condition, and MedicationRequest, but should not have access to DocumentReference or Communication. The purpose boundary determines why the AI is accessing data and ensures that access is constrained to that purpose.

SMART on FHIR scopes operationalize all three boundaries. System-level scopes (system/Patient.read) define the resource boundary. Launch contexts (patient/Observation.read) define the patient boundary. Application registration and scope negotiation define the purpose boundary — because the scopes an application is permitted to request are configured by the health system administrator during vendor registration, not by the vendor alone. This is a crucial point for healthcare managers: you control what scopes each AI application is allowed to request. The authorization server enforces that boundary. The AI vendor does not unilaterally decide what they can access.

Operationalizing appropriate access requires more than initial configuration. It requires ongoing review. AI systems evolve. Vendors push updates. New features request new data types. A governance process that reviews scope requests at each major version update and requires vendor documentation explaining why each scope is necessary for each feature is not bureaucratic overhead — it is the minimum standard of responsible oversight for systems that touch protected health information.

4.4.2 Preventing Misuse and Over-Permissioning

Over-permissioning is the most common OAuth failure mode in enterprise deployments, and healthcare is not exempt. It happens when applications request broad scopes because it is easier than scoping precisely, when administrators approve scope requests without scrutiny, and when access reviews are infrequent or nonexistent. The result is AI systems — and other applications — that have far more access to data than they need, creating unnecessary exposure, regulatory risk, and liability.

The principle of least privilege is the corrective. It holds that any system should have access to the minimum data required to perform its function, nothing more. Applied to AI in healthcare: a sepsis early warning system should have read access to vital signs, laboratory values, and nursing assessments — and nothing else. It should not have write access to any resource. It should not have access to psychiatric notes or reproductive health records. It should not have access to patients not currently on the monitored unit. These constraints are not just privacy protections — they are also AI quality controls, because access to irrelevant data can introduce noise that degrades model performance.

Misuse takes several forms in the AI context. Scope creep — gradual expansion of access beyond initial authorization — is the most common. Model exfiltration — using API access to extract patient data for purposes beyond the stated use case — is the most serious form of exfiltration. Inference attacks — using permitted access to infer data that is not directly accessible indirectly — are an emerging concern for AI systems. Healthcare managers should understand that preventing misuse is not just a matter of setting scopes at deployment; it requires monitoring token usage patterns, reviewing API call logs, and comparing actual data access against the stated purpose of the AI system.

Vendor contracts for AI systems should include explicit language about permitted data use, prohibitions on secondary use, requirements for access logging, and the right to audit. These contractual controls work in concert with the technical controls provided by OAuth. Neither alone is sufficient: technical controls without contractual accountability leave a vendor free to circumvent scoping through application-level data aggregation; contractual controls without technical enforcement leave a vendor capable of exceeding permitted access with no automatic check.

A practical checklist for preventing over-permissioning in AI deployments:

- Review the complete scope list requested by each AI application before granting authorization server registration.
- Require the vendor to document the functional need for each requested scope — do not accept blanket justifications.

- Configure the authorization server to limit which scopes each registered application may request, regardless of what the application asks for.
- Set short access token lifetimes for AI systems with continuous access, and configure refresh token rotation so that each refresh produces a new token and invalidates the old one.
- Conduct quarterly access reviews — not just of user accounts but of all registered AI applications and their current scope grants.
- Establish an alert for anomalous API call volumes from AI system tokens — sudden spikes may indicate misuse or a compromised application.
- Include a right-to-audit clause in every AI vendor contract, covering both technical access logs and internal data handling practices.

4.4.3 How OAuth and OIDC Enable Safe Automation

Healthcare AI systems frequently operate in automated, unattended modes — running predictive models overnight, generating population health reports, triggering alerts when patient conditions meet defined thresholds. These are backend processes with no human user approving each individual data access. This is where the SMART on FHIR backend services authorization profile becomes essential.

The backend services profile uses the OAuth 2.0 client credentials flow, adapted with a signed JWT assertion that the AI system generates and presents to the authorization server. The authorization server validates the JWT—

checking the signature, issuer, audience, and expiration—and issues an access token if the assertion is valid. No human user is in the loop during token issuance. But the trust relationship is established beforehand: the health system's authorization server has been configured to recognize the AI system's signing key as a trusted credential, scoped to specific resources and permissions.

This is what responsible-by-design looks like for automated AI: the automation is real, the speed is real, but the trust boundary is pre-established, auditable, and revocable. The AI system cannot unilaterally access new data types or patient populations — it can only operate within the scope boundary configured by the health system administrator. And every access event is logged against the AI system's registered identity, creating a complete audit trail.

Safe automation also depends on failure handling. When an AI system's access token expires, and it cannot obtain a new one — because the refresh token has been revoked, the client credentials have been rotated, or the authorization server is temporarily unavailable — the system should fail gracefully. It should not attempt to cache data locally while continuing to operate outside the authorized session. It should not attempt to use expired tokens. It should halt, log the failure, and alert the appropriate technical team. Building this fail-safe behavior into AI system requirements is a governance responsibility, not just an engineering preference.

Diagram 4.5 – Backend AI Authorization: The SMART Backend Services Flow

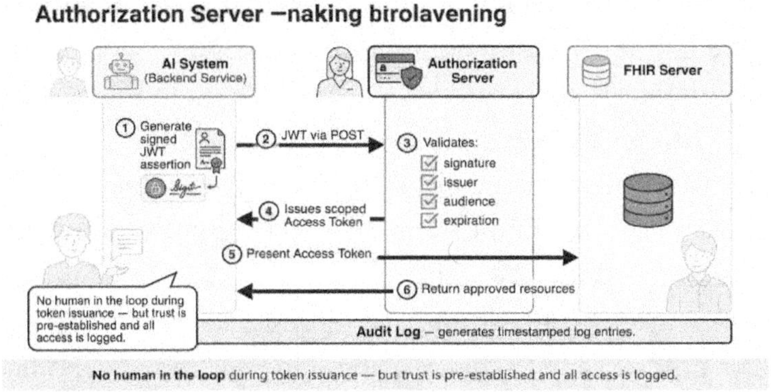

4.4.4 Building a Governance Model for AI Access

OAuth and OIDC provide the technical controls. They do not provide themselves. Building a governance model for AI access means assigning organizational accountability for every layer of the trust framework: who registers AI applications, who approves scope grants, who conducts access reviews, who responds to security incidents involving AI tokens, and who has the authority to revoke access when necessary.

In most health systems, this accountability is distributed across multiple teams — information security, compliance, IT operations, clinical informatics, and the legal or contracts team. The risk in distribution is diffusion: when everyone is a little bit responsible, no one is fully responsible. A governance model for AI access should designate a single accountable owner for the authorization server configuration, a defined approval process for new AI application registrations, and a named decision-maker with authority to revoke access in the event of a security incident.

The governance model should be documented and tested, not just designed. A tabletop exercise — 'a vendor AI system has been breached; what do we do in the next hour?' — will reveal gaps in revocation procedures, escalation paths, and communication protocols faster than any policy document. For healthcare managers leading AI deployments, investing time in this exercise before a production deployment is mission-aligned preparation. After a breach, it is not the time to discover that no one knows who has the authorization server admin credentials.

The governance model should also account for the lifecycle of AI systems, not just their deployment. What happens when a vendor contract ends? What happens when an AI system is deprecated and replaced by a newer version? What happens when the vendor is acquired, and the product is migrated to a new platform? Each of these events requires specific actions: revoking old application registrations, rotating credentials, reviewing scope grants for the new deployment, and updating audit trail configurations. Building these lifecycle events into your AI governance calendar — and assigning ownership for each — is the difference between a security posture that holds over time and one that erodes quietly between annual reviews.

A sequence for establishing AI application governance:

1. Define a formal AI application registration process that requires vendor documentation of intended use

case, data accessed, scope justification, and security practices before any authorization server registration is approved.

2. Assign an Authorization Server Administrator role with documented responsibilities for configuration, scope management, and incident response.

3. Configure the authorization server to enforce scope limits at the application level — each registered AI application should have a defined maximum scope that cannot be exceeded, regardless of what the application requests.

4. Establish access token and refresh token lifetimes that balance operational continuity with security — shorter is safer, and automation should handle token refresh, not human intervention.

5. Implement API call logging that captures, at minimum, the application identifier, the resource type accessed, the patient identifier (where applicable), the timestamp, and the token identifier.

6. Conduct quarterly access reviews covering all registered AI applications: are they still active, are their scopes still appropriate, are their API call volumes consistent with their stated use case?

7. Conduct an annual tabletop exercise simulating an AI application compromise, testing the revocation procedure and the incident response escalation path.

8. Document the AI application decommissioning procedure and execute it for every system that reaches end of life — including credential rotation for any shared infrastructure.

Diagram 4.6 – AI Access Governance Lifecycle

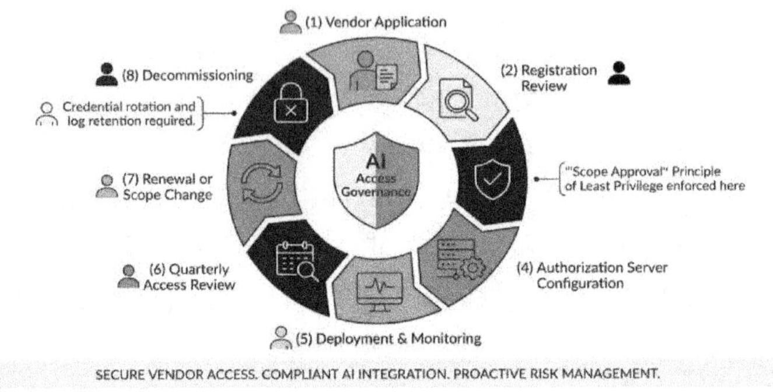

4.4.5 Federated Identity and Cross-Organizational Scenarios

Many healthcare AI deployments span organizational boundaries. A regional health information exchange may deploy an AI analytics platform that draws data from multiple member hospitals. A national payer may operate AI-driven care management tools that integrate with provider EHR systems across hundreds of organizations. An ACO may license a vendor AI platform that needs access to data from both its acute care partners and its post-acute network. In each scenario, a single identity provider is not sufficient — the AI system must operate within a federated identity model.

Federated identity means that multiple identity providers trust a common set of identity assertions, or trust a broker that translates between them. OpenID Connect supports federation through a model where a central identity broker aggregates identity assertions from multiple participating organizations and presents a unified identity layer to downstream applications. AI systems in federated environments receive identity tokens that may assert the user's home organization, their role within that organization, and their participation in a shared care network — all in a single, verifiable token.

For healthcare managers in multi-entity AI deployments, the governance complexity multiplies. Each participating organization must agree on which identity attributes are asserted, how roles are mapped across organizations, and which entity is the authoritative source for each attribute. These are not engineering decisions alone — they require policy agreement at the organizational level. The technical implementation is straightforward once the policy is settled; the policy alignment is where multi-organizational AI projects most commonly stall.

The pilot-to-production pathway for cross-organizational AI deployments should include an explicit identity federation design phase: identifying all participating identity providers, mapping the identity attributes required by the AI system, establishing a federation policy agreement, and testing the complete authentication and authorization flow with each participating organization before any clinical data is processed. This phase is often underestimated and underbudgeted, and its omission is one of the most common causes of production delays in health system AI integration projects.

4.5 Takeaway

OAuth 2.0 and OpenID Connect are not features of healthcare AI — they are the foundation on which it stands.

Without them, every AI system that touches clinical data is operating on unverified trust, with uncontrolled access, and without an auditable record of who accessed what and why. With them, AI systems operate within defined boundaries, on behalf of verified identities, with access grants that can be reviewed, constrained, and revoked.

The healthcare manager's role in this framework is not to implement the protocols — that is, engineering work. The manager's role is to establish the governance model that ensures the protocols are configured correctly, monitored continuously, and enforced consistently. That means owning the application registration process, requiring scope justifications, conducting access reviews, and ensuring that the revocation procedure is tested before it is needed.

Three principles should anchor every AI access governance decision. First, least privilege: AI systems should access only the data they need for their stated purpose, and that constraint should be enforced technically, not just contractually. Second, auditability: every data access event by an AI system should produce a log entry that can be reviewed, reported, and investigated. Third, revocability: the ability to cut off any AI system's access within minutes should be a standing operational capability, not an emergency measure that requires days to execute.

The trust that OAuth and OIDC establish is not automatic. It is constructed through careful configuration, active governance, and deliberate organizational alignment. Healthcare leaders who invest in that construction are not just protecting their organizations from risk. They are building the infrastructure that enables safe, scalable, mission-aligned AI adoption. Every AI system you deploy on a solid trust framework is one more step from pilot to production, and one more capability available to the clinicians and patients who depend on your organization to get this right.

In the next chapter, we examine SMART on FHIR in depth — how it combines OAuth, OIDC, and FHIR into a unified application launch and authorization framework, and what it means practically for deploying clinical AI applications within and across health systems.

5 Healthcare Interoperability: The Bigger Picture

Interoperability is one of those words that means everything and nothing at once. Ask ten healthcare leaders what it means, and you will get ten different answers — ranging from "systems talking to each other" to "the ability to share lab results across hospitals" to "our EHR vendor's API strategy." All of those answers are partially right. None of them is sufficient.

The narrow definition — technical data exchange — is where most organizations focus their energy, and most projects stall. Real interoperability is something broader. It is the capacity of people, processes, and systems across different organizations to work together toward a common patient care goal, without friction that degrades quality, safety, or efficiency. When that capacity is missing, the consequences are not abstract. Patients are harmed. Clinicians burn out. Organizations absorb costs that should never exist.

FHIR is a critical enabler of interoperability, but it is not interoperability itself. Managers who treat FHIR as a silver bullet — deploy the API, declare victory, move on — will find themselves back at the same table in two years, dealing with the same problems wearing slightly different names. The value of FHIR is unlocked only when it is embedded in a broader strategy that addresses governance, semantics, workflow, and organizational culture.

This chapter builds that broader picture. It defines the four layers of interoperability that every healthcare manager needs to understand. It locates FHIR precisely within that landscape — including its real strengths and its real limits. It examines the actual cost of interoperability failures. And it translates all of that into a strategy framework that managers can bring to budget conversations, vendor negotiations, and cross-functional planning sessions.

The goal is operational clarity: a way of thinking about interoperability that is mission-aligned, grounded in evidence, and useful for making decisions — not just discussing them.

5.1 The Four Layers of Interoperability

Healthcare interoperability does not fail in one place. It fails across multiple layers simultaneously, which is why single-point fixes — a new API here, a new integration engine there — rarely produce durable results. The industry has converged on a four-layer model, adapted from HIMSS and other standards bodies, that provides managers with a reliable map of where problems lie and where investment should go.

Think of the four layers as a stack. Each layer depends on the one below it. Technical interoperability is the foundation. Semantic interoperability sits on top of it. Workflow interoperability builds on semantics. Governance binds them all together and determines whether the system as a whole behaves reliably over time. A gap at any layer undermines every layer above it.

Diagram 5.1 – The Four Layers of Interoperability

EXECUTIVE STRUCTURE INFOGRAPHIC

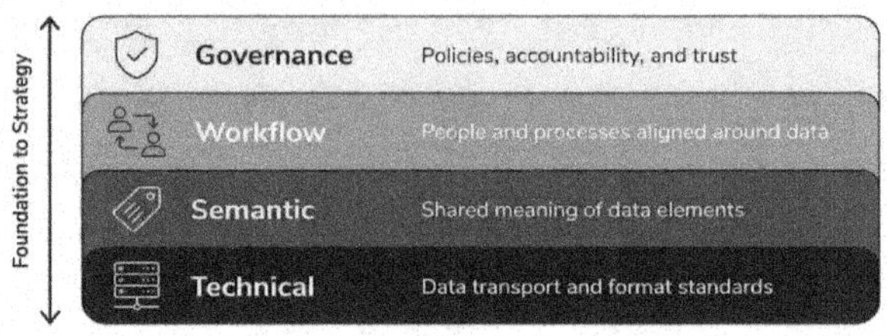

5.1.1 Technical

Technical interoperability is the ability to transmit data from one system to another using agreed-upon formats and transport protocols. This is the infrastructure layer — the plumbing. It asks: Can these two systems physically exchange a message?

For most of healthcare's history, the answer was: barely. HL7 v2 messages — the aging workhorse standard introduced in the 1980s — enabled rudimentary exchange, but with implementation variability so wide that two organizations both claiming HL7 compliance could often not parse each other's messages without custom mapping. FHIR modernizes this layer by standardizing not just the format (JSON or XML) but the transport mechanism (RESTful HTTP) and the resource structures. When two systems speak FHIR, they share a common grammar at the technical layer that is far more consistent than that of HL7 v2.

Managers need to understand what technical interoperability does and does not guarantee. It guarantees that a message arrives intact and in a form the receiving system can parse. It does not guarantee that the receiving system interprets the message correctly, routes it to the right workflow, or uses it to improve care. Those are problems for the layers above.

Common technical interoperability decisions managers face include: which FHIR version to mandate in vendor contracts (R4 is the current baseline; R4B and R5 add specific capabilities), whether to require certified SMART on FHIR endpoints, and whether integration infrastructure is centralized (a single integration engine or platform) or federated (each system manages its own connections). These are not purely technical choices — they have downstream effects on cost, security surface area, and the organization's ability to adopt new tools without custom integration work.

5.1.2 Semantic

Semantic interoperability is the ability of two systems to share not just data but shared meaning. A lab result of "glucose: 110" means nothing unless both systems agree on the units (mg/dL or mmol/L?), the context (fasting? random?), the reference range, and what code system identifies the test. Semantic interoperability answers the question: Do we both mean the same thing by this data element?

This is where many FHIR implementations quietly break down. The technical exchange works. The message arrives. But the receiving system maps the incoming code to a local concept that is subtly different; the alert fires incorrectly; the AI model trained on one code system makes an inference using data coded in another; and the clinical decision is degraded. This failure is invisible in most monitoring dashboards because the system does not know it does not know.

The vocabulary standards that underpin semantic interoperability are well-established: SNOMED CT for clinical terminology, LOINC for lab results and clinical observations, RxNorm for medications, ICD-10-CM for diagnoses, and CPT for procedures. The US Core Implementation Guide — built on FHIR R4 — mandates specific value sets and terminologies for each resource type. Compliance with US Core is the minimum bar for semantic interoperability in US healthcare contexts.

The manager's challenge is not knowing which vocabulary standards exist — it is building processes that enforce their use consistently. Vendors frequently implement local code systems that approximate standard terminologies without exactly matching them. Every local code system is a future translation cost. Contracts should require compliance with standard terminology as a condition of acceptance testing, not as an aspiration in a statement of work.

Semantic interoperability also matters directly for AI. A model trained on LOINC-coded data will produce unreliable results when applied to data coded in a proprietary local system. Organizations that have invested in semantic standardization find that AI produces value faster and with fewer corrective cycles. Those who skipped semantic cleanup discover it when their AI model cannot be trusted in production.

5.1.3 Workflow

Workflow interoperability is the capacity of people and processes in different organizations to collaborate effectively, supported by the data flowing between their systems. It answers the question: even if data is shared correctly, do the processes on either end act on it in a coordinated, timely, and clinically appropriate way?

Workflow interoperability is where the human dimension of data exchange becomes visible. A discharge summary that arrives at the primary care physician's EHR within seconds of discharge is a technical and semantic success. If the PCP's workflow does not include a structured process for reviewing incoming transition-of-care documents, or if the document arrives as an unformatted PDF attachment buried in an inbox alongside 200 other messages, the technical achievement produced no clinical benefit.

CDS Hooks — a FHIR-adjacent standard — is one of the most important tools for workflow interoperability. It allows external services to inject clinical decision support at specific points in a clinician's EHR workflow: when a patient's chart is opened, when a medication is ordered, and when a referral is placed. CDS Hooks does not just push data; it inserts guidance when a clinician can act on it. This is workflow-level impact in the most direct sense.

Managers assessing workflow interoperability should map the clinical workflows that depend on external data —

transitions of care, referral management, medication reconciliation, and care gap closure — and trace them back to the data requirements for each. The question is not "does data flow?" but "does data flow to the right person, in the right format, at the right moment in their workflow, with the right context to act on it?" That is a much harder standard to meet, and it is the one that actually affects patient outcomes.

Workflow interoperability decisions often sit at the intersection of IT and clinical operations. They require clinician engagement, change management, and process design — not just technical configuration. Projects that treat workflow as an afterthought to technical integration consistently underperform. Projects that design for workflow impact first, then work backward to the technical requirements, consistently outperform.

Diagram 5.1b – CDS Hooks Workflow Integration

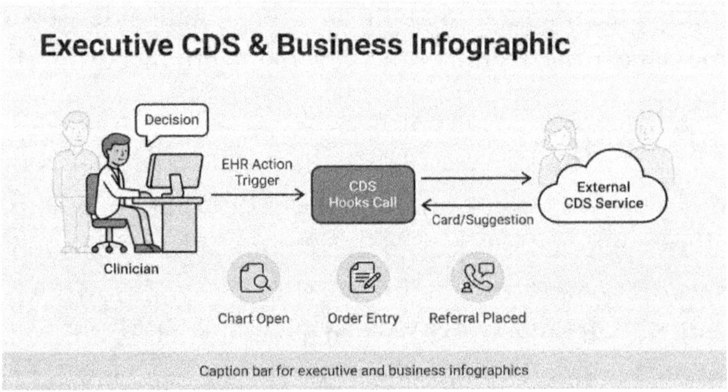

5.1.4 Governance

Governance interoperability is the layer of policies, agreements, accountability structures, and trust frameworks

that allow organizations to share data across legal, organizational, and jurisdictional boundaries. It answers the question: Do we have the authority, the agreements, and the accountability structures to share and use this data responsibly?

This layer is invisible when it works and catastrophic when it does not. An organization can have flawless technical and semantic interoperability with a partner network and still be unable to share data if the Business Associate Agreements are not in place, the data use agreements do not cover the specific intended use, or the consent framework does not permit secondary use for the purpose in question. Governance failures pose legal risks, regulatory exposure, and organizational liability.

The governance layer has become more complex, not simpler, over the past decade. The 21st Century Cures Act mandated open APIs and prohibited information blocking — but it also introduced a complex framework of permitted exceptions. State-level privacy laws increasingly diverge from HIPAA, creating compliance complexity for multi-state organizations. AI adds new governance requirements: data provenance, model validation, bias auditing, and explainability obligations that did not exist when most health data governance frameworks were designed.

For managers, governance interoperability requires proactive investment in legal infrastructure, not just technical infrastructure. Data sharing agreements should be templated, reviewed, and maintained as living documents — not treated as one-time legal events. The organization's data governance function needs a seat at the table for every significant interoperability initiative, not a cameo appearance at the sign-off stage.

Responsible by design means building governance requirements into the architecture of interoperability programs from the start. This includes patient consent management, data minimization principles, audit logging, and clear delineation of who is accountable for data quality, data security, and appropriate use at every point in the data flow. Organizations that embed governance into the design of their interoperability architecture handle audits and regulatory inquiries with confidence. Those who treat governance as an afterthought face them with fear.

5.2 Where FHIR Fits In

FHIR is the most significant advance in healthcare data standards in a generation. It has done more to normalize the technical infrastructure of healthcare data exchange than any standard before it. Understanding where FHIR excels and where it reaches its limits is essential for managers who need to make sound decisions about where to invest, what to demand from vendors, and what problems cannot be solved by FHIR alone.

FHIR's adoption trajectory is no longer a question of "if" — it is a question of depth and consistency. The CMS Interoperability and Patient Access Final Rule mandates FHIR R4 APIs for Medicare and Medicaid payers. The ONC 21st Century Cures Act rule requires certified health IT to expose FHIR-based APIs. Virtually every major EHR vendor has a FHIR API roadmap. The standard is embedded in federal regulatory requirements in a way that makes avoiding it increasingly untenable. The manager's job is not to decide whether to adopt FHIR but to decide how to adopt it strategically.

Diagram 5.2 – FHIR's Position in the Interoperability Stack

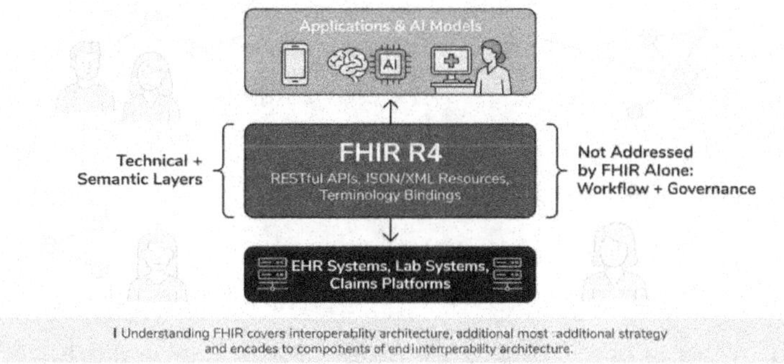

5.2.1 Strengths and Limitations

FHIR's core strengths are well documented and genuinely substantial. At the technical layer, FHIR's RESTful architecture means that any developer familiar with web APIs can work with healthcare data without specialized EDI or HL7 v2 expertise. This dramatically expands the talent pool available for healthcare application development and reduces integration costs. The resource model — patient, encounter, observation, condition, medication — provides a coherent domain model that maps reasonably well to clinical reality. Versioning, error handling, and search capabilities are built into the standard in ways that make them predictable and portable.

At the semantic layer, FHIR's Implementation Guides (IGs) do substantial work. The US Core Implementation Guide specifies required data elements, terminology bindings, and conformance expectations for common clinical data classes. The USCDI (United States Core Data for Interoperability) standard defines the minimum data elements that certified EHRs must expose. When a vendor's FHIR API is certified to US Core, a manager can have reasonable confidence that the most commonly needed clinical data elements will be available, coded consistently, and accessible via standard API calls. That is a genuine advance over the pre-FHIR world.

FHIR also provides the authentication and authorization layer — through SMART on FHIR — that enables secure, granular access control. An application that is authorized via SMART on FHIR can request only the data scopes it needs, authenticated through the EHR's OAuth 2.0 infrastructure. This is mission-aligned security: the access model matches the clinical context, and audit trails are built in.

FHIR's limitations are equally real, and managers must internalize them. First, FHIR is a standard for data exchange, not data quality. If the data in the source system is incomplete, miscoded, or clinically inaccurate, FHIR will efficiently transmit that bad data while remaining fully conformant. "FHIR-compliant" does not mean "clinically reliable." Data quality programs are not replaceable by interoperability standards.

Second, FHIR implementation variability remains a significant problem in real-world settings. The standard is permissive in many areas — optional fields, extensions that can be used without limit, value sets that can be expanded locally. Two EHR vendors can both claim FHIR R4 compliance while implementing the standard in ways that require custom mapping to achieve semantic interoperability between them. Implementation Guides narrow this variability, but they do not eliminate it. Conformance testing against a published IG — not just vendor claims of compliance — is the only reliable check.

Third, FHIR does not fully capture the complexity of clinical documents. Structured data — discrete lab values, coded diagnoses, medication lists — maps cleanly to FHIR resources. Unstructured or semi-structured clinical content — physician notes, radiology reports, complex care plans with embedded reasoning — is represented poorly or inconsistently in FHIR. The clinical note remains one of the richest sources of clinical information and one of the least accessible via FHIR APIs.

Fourth, FHIR says nothing about what to do with data once it is exchanged. The workflow and governance layers are entirely outside its scope. An organization with a perfect FHIR implementation can still have broken workflows, no data use agreements, and no accountability for data quality. FHIR enables the plumbing; it does not build the house.

5.2.2 Why FHIR Is Necessary but Not Sufficient

The phrase "necessary but not sufficient" is the most accurate description of FHIR's role in healthcare interoperability. Necessary because: without a modern, standardized data exchange layer, every interoperability effort requires bespoke integration work that is expensive, fragile, and impossible to scale. The operational and financial cost of a pre-FHIR integration environment is enormous — most organizations that have done the math have found it runs to millions of dollars annually in integration maintenance alone.

Not sufficient because: the problems that matter most to patients and healthcare organizations — care coordination failures, AI model reliability, regulatory compliance, organizational accountability for data — require investment across all four interoperability layers. A manager who allocates the entire interoperability budget to FHIR API implementation and treats semantic, workflow, and governance work as optional will eventually face a clinical or regulatory event that proves the investment was incomplete.

The practical implication for budget allocation is clear: FHIR API compliance is a prerequisite, not a destination. Once the technical foundation is in place, the highest-value investments are typically in semantic standardization (vocabulary governance, terminology mapping), workflow redesign (clinical workflow analysis, CDS integration, alert optimization), and governance infrastructure (data use agreements, consent management, audit capability). These are not glamorous investments. They do not generate vendor press releases or conference presentations. But they are the investments that convert technical interoperability into measurable clinical and operational outcomes.

Managers who communicate this framing to leadership and boards — FHIR as necessary foundation, not complete solution — set realistic expectations and secure ongoing investment for the full program. Those who present FHIR as the solution to interoperability tend to find their credibility questioned when outcomes do not follow from API deployment.

5.3 Interoperability Failures and Their Impact

Interoperability failures are not theoretical. They have documented consequences for patient safety, operational performance, financial sustainability, and patient experience. Managers who frame interoperability investment in terms of avoided failure — not just enabled capability — tend to win more budget battles and sustain more organizational commitment to the work.

The body of evidence on interoperability failures is large and sobering. A 2020 CAQH report estimated that the healthcare industry wastes approximately $13.3 billion annually on administrative transactions that could be automated through interoperability. The Office of the National Coordinator for Health Information Technology has documented hundreds of cases of information blocking — deliberate suppression of interoperability — that required regulatory action. Patient safety research consistently identifies information fragmentation as a contributing factor in adverse events across care settings.

Diagram 5.3 – The Cost Cascade of Interoperability Failure

Analytical diagram of the consequences of 'Interoperability Failure' in healthcare

5.3.1 Patient Safety Risks

The most consequential cost of interoperability failure is harm to patients. The mechanisms are numerous and well understood. When a patient's full medication list is not available at the point of prescribing, drug-drug interactions go undetected. When an allergist's documentation does not reach the emergency department, known allergens are administered. When a specialist's findings are not integrated into the primary care record, diagnoses are missed and chronic conditions escalate without appropriate management.

Transitions of care are the highest-risk moments. Hospital discharge to post-acute care, emergency department to primary care follow-up, specialist consultation back to the referring physician — these handoffs are where data fragmentation most reliably produces adverse outcomes. A 2019 study published in the Journal of the American Medical Informatics Association found that incomplete medication information at hospital discharge was associated with a significantly elevated risk of readmission within 30 days.

The data was available somewhere in the system; it just was not available in the right system at the right time.

Duplicate diagnostic testing is another safety risk that is often miscategorized as merely a cost problem. A patient who has a CT scan repeated because the results from a prior scan at a different institution are inaccessible is exposed to unnecessary radiation, delayed diagnosis, and the opportunity costs of wasted clinical time. The financial waste is real, but the clinical risk is the primary concern.

For managers, patient safety risk is the most powerful argument for interoperability investment. It is also the argument most likely to resonate with clinical leadership, patient safety committees, and governing boards. Framing interoperability failures as a patient safety issue — not just an IT infrastructure issue — positions the program correctly and tends to unlock resources that IT budget cycles alone cannot reach.

5.3.2 Operational Inefficiencies

Operational inefficiency from interoperability failure is pervasive and largely invisible — until it is measured. The most common manifestation is manual data reconciliation: clinical staff spending time re-entering data that exists in another system but cannot be accessed programmatically. Medication reconciliation performed by nursing staff at admission is among the most labor-intensive examples: in organizations without effective interoperability, this process can take 30-60 minutes per patient and is still error-prone because the data sources — patient self-report, pharmacy records, specialist notes — are not integrated.

Prior authorization is a canonical example of operational inefficiency driven by interoperability failure. The payer needs clinical documentation to adjudicate the authorization request. The provider's staff must locate, format, and transmit that documentation manually — typically by fax or portal upload — because the payer's system cannot directly query the provider's EHR. This process consumes an estimated 14.9 hours per physician per week across the healthcare workforce, according to the American Medical Association. Fully automated prior authorization via FHIR Da Vinci APIs could eliminate most of that burden. The technology exists. The adoption lag is a governance and workflow problem, not a technical one.

Referral management is another high-volume process where interoperability failure drives operational waste. When a referring provider sends a referral to a specialist, the specialist practice typically cannot access the referring provider's clinical documentation without a phone call or fax request. The result is scheduling delays, appointment cancellations when documentation is missing, and duplicate workups. An integrated referral management workflow — where the specialist receives structured FHIR data at the time of referral, including the relevant clinical context, current medications, and prior workup results — collapses this waste to near zero.

Managers should quantify these operational inefficiencies before building the business case for interoperability investment. Time-motion studies in target workflows, analysis of rework rates, and measurement of cycle times for key administrative processes provide the data needed to build an ROI argument. In most organizations, the operational savings from full interoperability in even two or three high-volume workflows pay for the technical investment within 18-24 months.

5.3.3 Financial Waste

The financial cost of interoperability failure is both direct and indirect. The direct costs include duplicate testing (estimated at $8-9 billion annually across the US healthcare system), administrative transaction inefficiency ($13.3 billion in avoidable manual processing per the CAQH analysis), preventable readmissions attributable in part to inadequate care transition data, and the maintenance cost of point-to-point integrations built to work around the absence of standard APIs.

The indirect costs are harder to quantify but often larger. An AI-powered care management program that fails to achieve its projected outcomes because the data it relies on is fragmented or inconsistently coded represents a significant loss of investment — not just in the AI itself but also in the organizational change management, clinical workflow redesign, and vendor contracts that surround it. Failed AI programs in healthcare are frequently diagnosed as AI problems; in many cases, they are interoperability problems in disguise.

For value-based care organizations, interoperability failure has direct financial consequences under risk contracts. A population health program that cannot identify high-risk patients because its risk stratification model depends on claims data alone — without the clinical data available in partner EHRs — will systematically miss patients who need intervention. Each missed high-risk patient who experiences a preventable hospitalization represents a direct financial loss under a capitated or shared-savings contract.

The financial argument for interoperability is strong, but it requires discipline to make correctly. The savings are real but distributed — across clinical staff time, administrative labor, duplicate testing, and risk contract performance. Building a credible financial case requires aggregating across these categories rather than relying on any single line item. Managers who do this analysis rigorously find that interoperability investment generates positive returns. Managers who make qualitative claims about efficiency without quantitative backing find those claims challenged at budget time.

5.3.4 Poor Patient Experience

The patient experience cost of interoperability failure is underappreciated in most business cases. Patients experience interoperability failure directly: being asked the same questions by multiple providers, having to physically carry paper records between institutions, being unable to access their own health data in a usable form, and receiving duplicated or contradictory instructions from disconnected care teams. These experiences damage trust in the healthcare system and contribute to disengagement from care.

Patient-facing FHIR applications — enabled by the 21st Century Cures Act's Patient Access API requirements — represent a significant opportunity to close this gap. Patients have a legal right to their health data in FHIR format, accessible via third-party apps of their choosing. When organizations implement robust Patient Access APIs and support patient-mediated data sharing, patients can bring their complete record to any provider encounter. The data burden shifts from the provider network to the patient — who is often more motivated to manage their own data than any individual provider in the care team.

HCAHPS scores and patient satisfaction metrics are increasingly sensitive to care coordination. Patients who experience seamless information sharing — where their new specialist already has their records, where their primary care physician is notified of an ED visit in real time, where their discharge instructions match their outpatient medication list — rate their care significantly higher than patients who experience fragmented communication. As payer contracting increasingly incorporates patient experience metrics, the business case for patient-centered interoperability strengthens.

For managers, the patient experience dimension of interoperability connects to consumer strategy, market positioning, and long-term patient retention. Organizations that lead on interoperability — where patients can access their data, where care coordination is visible and reliable — differentiate themselves in ways that matter to consumers making choices between competing health systems. In competitive markets, interoperability is not just an operational investment; it is a market strategy.

5.4 Interoperability as Strategy

Most organizations treat interoperability as a compliance activity or an infrastructure project. The organizations leading in healthcare — in outcomes, in AI capability, in operational efficiency, and in market position — treat it as a strategic asset. The difference in approach is not technical. It is organizational: how leadership thinks about data, how governance is structured, and how cross-functional teams are aligned around the goal of making data useful across the enterprise.

Interoperability strategy means making deliberate decisions about which data flows matter most, building the organizational capability to maintain and improve those flows over time, and creating accountability structures that ensure data quality is everyone's responsibility — not just IT's problem. It means thinking about the data landscape the way a supply chain manager thinks about logistics: with visibility into every node, clear ownership of every handoff, and explicit standards for performance at every stage.

Diagram 5.4 – Interoperability as a Strategic Asset

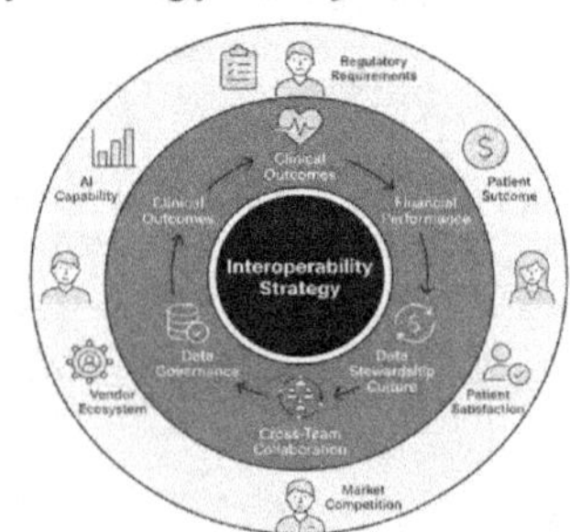

:Caption Text:

5.4.1 Data Governance

Data governance is the set of policies, processes, roles, and accountabilities that determine how data is created, maintained, used, and retired across an organization. In the context of interoperability, governance answers the questions that technology cannot: Who owns this data? Who is accountable for its quality? What can it be used for? How long can it be retained? Who approves new data sharing agreements?

Effective data governance for interoperability requires a defined organizational structure. The minimum viable structure for most healthcare organizations includes: a Chief Data Officer or equivalent executive accountable for enterprise data strategy; a Data Governance Council with representation from clinical operations, IT, legal and compliance, finance, and — critically — clinical informatics; a defined process for reviewing and approving data sharing agreements; and a data dictionary or master data management function that maintains authoritative definitions of key data elements.

1. Establish executive accountability: assign a senior leader — CDO, CIO, or CMIO — as the accountable executive for enterprise data governance, with explicit authority over data sharing decisions.
2. Form a Data Governance Council: convene representatives from clinical, operational, legal, IT, and analytics functions. Meet at least quarterly; more frequently during active interoperability programs.
3. Build a data inventory: catalog all data assets, their sources, their quality characteristics, and the

workflows that depend on them. This inventory is the map that enables governance decisions.

4. Template data sharing agreements: develop and maintain standard BAA and data use agreement templates reviewed by legal counsel. Reduce approval cycle time from months to weeks.
5. Implement terminology governance: assign ownership of vocabulary standards management. Require standard terminology as a contractual condition for all new system acquisitions.
6. Define data quality standards: establish measurable data quality targets — completeness rates, coding accuracy rates, timeliness standards — for each critical data domain. Report these metrics to the Governance Council quarterly.

Data governance is not a project — it is a function. Organizations that treat it as a one-time initiative — build a framework, declare success, move on — consistently find that the framework deteriorates as systems, vendors, and use cases evolve. The governance function needs ongoing investment: staff, tooling, and executive attention. The return on that investment is an organization that can execute new interoperability programs faster, with fewer legal obstacles, and with higher confidence in data quality at launch.

AI amplifies the importance of data governance because AI systems can act on data at scale. An AI model that operates on poorly governed data — data that is incomplete, mis-coded, or subject to data-use restrictions — can cause harm at a scale and speed that manual processes cannot match. Every AI initiative requires a data governance review before deployment to production. Governance built proactively enables AI programs to move from pilot to production on predictable timelines. Governance built reactively, after problems emerge, is far more expensive and disruptive.

5.4.2 Cross-Team Collaboration

Interoperability programs fail for organizational reasons at least as often as they fail for technical reasons. The most common organizational failure mode is functional siloing: IT owns the technical infrastructure, clinical informatics owns the data definitions, legal owns the agreements, clinical operations owns the workflows, and none of these groups communicates with sufficient frequency or shared purpose to produce a coherent result. Each team optimizes for its own objectives. The patient experience and operational outcomes that require all of them to work together never get fully owned.

Cross-team collaboration in interoperability requires structural solutions, not just cultural encouragement. The most effective structural solution is a dedicated interoperability program team — not a committee — with a program manager, technical leads, clinical informatics representation, and direct access to legal and compliance resources. This team owns the interoperability roadmap, manages vendor relationships, tracks progress against data quality targets, and escalates governance decisions to the Data Governance Council.

Workflow-level impact is achieved when IT and clinical operations solve problems together rather than sequentially. An interoperability program that delivers a technically correct data feed to a clinical workflow that was not redesigned to receive it will produce no clinical benefit. A program that co-designs the data feed and the receiving workflow — with clinicians and IT working side by side — produces outcomes that neither could achieve alone. This is the standard for cross-team collaboration in serious interoperability programs.

Vendor relationships are also a cross-team responsibility. EHR vendors, integration platform vendors, and application vendors all have commercial interests in how interoperability is implemented. Clinical leaders need to understand what they need from vendors to support their workflows. Legal and compliance need to understand vendor data handling practices. IT needs to evaluate vendor technical implementations against published standards. When these perspectives are synthesized, vendor negotiations are sharper and vendor performance expectations are more enforceable.

Diagram 5.5 – Cross-Functional Interoperability Team Structure

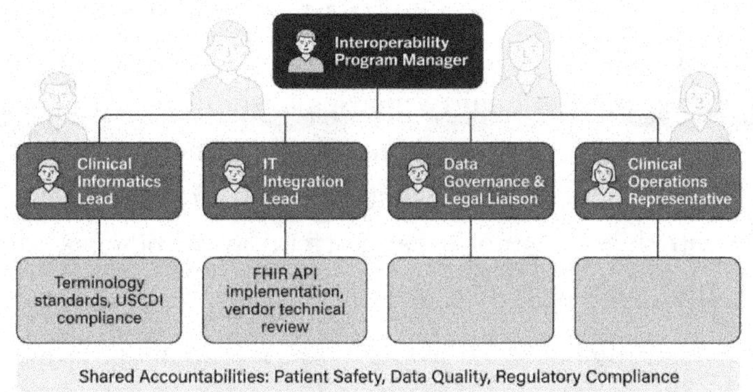

5.4.3 Building a Culture of Data Stewardship

Technology and governance structures are necessary but not sufficient conditions for successful interoperability. The organizations that sustain high-quality interoperability over time — through leadership transitions, vendor changes, and regulatory evolution — share a cultural characteristic: they treat data as a shared organizational asset that everyone is responsible for, not as IT's problem or the data team's problem.

Data stewardship culture starts with leadership behavior. When the CMO asks about data quality in clinical reviews, when the CFO treats data completeness as a financial risk, when the CEO includes interoperability progress in board reporting, clinicians and operational staff notice. The message that data quality matters is more powerful when it comes from clinical and operational leadership than from IT. IT can build the systems; only clinical and operational leaders can build the culture.

Concrete stewardship practices make culture tangible. These include designated data stewards for each major clinical domain — individuals responsible for monitoring data quality in their area and escalating problems; regular reporting of data quality metrics to clinical and operational leaders; and clear escalation paths when a data quality problem is detected that could affect patient care. Stewardship does not mean perfection; it means systematic attention and accountability.

Training is a stewardship tool that is consistently underinvested. Clinicians and administrative staff who understand why accurate, complete, and consistently coded data matters — who understand that their documentation decisions affect the next provider in the care continuum, the quality of AI-generated alerts, and the organization's risk contract performance — make better documentation decisions. Training that connects individual documentation behavior to patient outcomes and organizational performance is far more effective than training that focuses on system navigation.

Recognition and accountability complete the culture equation. Organizations that publicly recognize excellent documentation practices, that include data quality metrics in performance evaluations, and that investigate data quality failures with the same seriousness as other operational failures — these organizations build stewardship cultures that sustain interoperability programs through the inevitable challenges of complex healthcare environments. Those who treat data quality as a silent expectation, with no positive recognition and no real accountability for failure, find that the expectation is consistently unmet.

The connection between data stewardship culture and AI is direct and non-negotiable. AI models learn from data. AI models are validated using data. AI models are monitored using data. An organization that has built a culture of data stewardship has a vendor-independent AI foundation. An organization that has not built that culture will find its AI initiatives consistently underperforming against their projections — not because the AI is weak, but because the data it depends on is not managed with the discipline that AI requires.

Diagram 5.6 – Data Stewardship Culture Framework

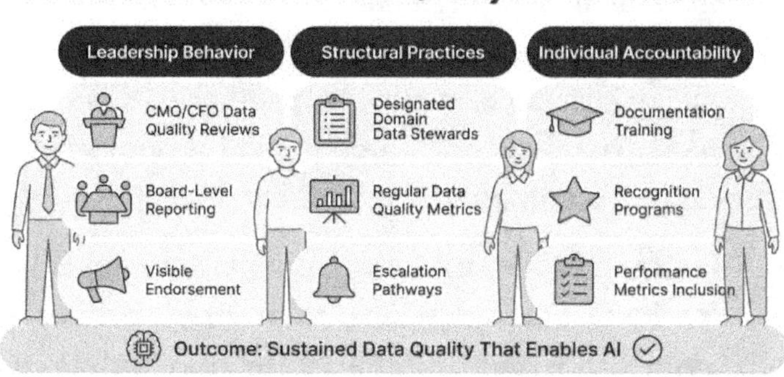

5.5 Takeaway

Healthcare interoperability is a four-layer challenge: technical, semantic, workflow, and governance. Solving one layer does not solve the others. Most organizations have underinvested in semantic standardization, workflow redesign, and governance infrastructure relative to their investment in technical integration. That imbalance is the root cause of most interoperability program failures and most AI implementation disappointments.

FHIR addresses the technical and, through Implementation Guides, much of the semantic layer. That is a significant advance. But FHIR's value is realized only when the organization has invested across all four layers. The pilot-to-production pathway for any interoperability-dependent program — including AI — requires governance infrastructure, workflow redesign, and data quality management alongside FHIR API implementation.

The failures of interoperability — patient safety events, operational waste, financial drag, poor patient experience — are not inevitable. They are the predictable outcomes of underinvestment and fragmented strategy. The organizations that treat interoperability as a strategic asset, build governance structures with real authority, and cultivate cultures of data stewardship consistently outperform those that treat it as a compliance exercise.

Managers who understand the four-layer model, who are honest about where their organization sits on each layer, and who build roadmaps that address all four layers over a realistic timeline are the managers who will successfully deploy AI, improve outcomes, and build the operational clarity that healthcare organizations urgently need.

The next chapter examines how FHIR-enabled data exchange creates the specific data conditions that AI models require — and what managers must build to convert interoperability infrastructure into AI-ready data pipelines.

6 How AI Learns from FHIR: Turning Data Into Intelligence

Every AI system in healthcare is only as good as the data it was trained on. That statement sounds simple, but its implications reach into every corner of a health system's operations. A model trained on incomplete records will produce incomplete predictions. A model trained on inconsistently coded diagnoses will generate inconsistent recommendations. And a model trained without a clear picture of who the patient actually is — comorbidities, medications, social history, recent labs — will operate with blind spots that can translate directly into patient harm.

FHIR changes this equation. The HL7 FHIR standard was not designed with AI in mind, but it turns out to be one of the most useful infrastructure investments a health system can make for its AI ambitions. When clinical, administrative, and operational data are structured consistently across systems, AI models have something they rarely get in healthcare: a reliable, machine-readable view of the patient journey. That reliability is what separates pilot projects that never scale from production AI that delivers workflow-level impact, quarter after quarter.

This chapter is for the decision-maker who needs to understand not just what FHIR does for interoperability, but what it does for intelligence. It covers how structured data shapes model quality, which AI techniques benefit most directly from FHIR's design, where bias and safety risks concentrate, and how standardized data creates a pilot-to-production pathway that proprietary or siloed data cannot support. The goal is operational clarity — the kind that lets a manager make a confident investment decision.

6.1 Why Structured Data Matters

Before any AI model makes a prediction, it must ingest data. Before it ingests data, that data must be findable, readable, and interpretable by a machine. In most health systems, this prerequisite alone stops AI initiatives before they start. Data sits in narrative notes, proprietary database schemas, disconnected departmental systems, and legacy formats that no modern API can reach. The gap between 'we have the data' and 'the AI can use the data' is often the longest and most expensive part of any AI project.

Structured data solves this. When clinical information is organized into discrete, consistently labeled fields — when a lab result is not just text in a note but a quantified observation with a LOINC code, a timestamp, a patient reference, and a reference range — the machine can work with it directly. No additional extraction step, no natural language processing to guess what the number means, no custom pipeline to reconcile the difference between how System A labels hemoglobin and how System B does. The data arrives AI-ready.

6.1.1 Garbage In, Garbage Out

The phrase 'garbage in, garbage out' is old enough to feel clichéd, but it remains the single most accurate description of what happens when AI models are trained on poor-quality data. In healthcare, garbage takes many forms.

It can be missing values — patients who appear in the emergency department record but whose outpatient medication history was never imported. It can be inconsistent coding — the same condition documented as ICD-10-CM E11.9 in one system and as free text 'Type 2 DM' in another, creating what looks like two different patient populations when it is actually one. It can be duplicated records, stale demographics, or fields that were designed for billing rather than clinical accuracy.

Each of these defects propagates through training. A model that sees incomplete medication lists will underestimate polypharmacy risk. A model trained on inconsistently coded diagnoses will assign different risk scores to clinically identical patients depending on which system documented their care. A model built on billing-driven data will optimize for what gets coded, not for what actually happened clinically. The downstream result is a system that may perform well on benchmark datasets but fails in production — precisely because production data reflects the same underlying inconsistencies the model was trained on.

This is where FHIR's design philosophy pays dividends. The standard enforces structure at the point of data definition, not at the point of AI consumption. When a health system has implemented FHIR-compliant APIs across its major platforms, lab results come in the Observation resource format, medications in the MedicationRequest format, and diagnoses in the Condition resource. Each carries mandatory and optional fields that create a consistent shape. That shape allows AI engineers to write feature-extraction pipelines that work across systems rather than for a single system at a time.

For the manager, this has a concrete financial implication. Data preparation — cleaning, normalizing, reconciling, transforming — typically accounts for 60 to 80 percent of the labor in an AI project. Organizations that have invested in FHIR infrastructure significantly compress this phase. They are not eliminating data quality work, but they are shifting it from custom, project-by-project transformation into standardized, reusable pipeline components. That shift compounds: the second AI model costs less to build than the first, and the fifth less than the second.

Diagram 6.1 – The Data Quality Chain: From Source to AI Decision

Executive/Business Infographic flora
Fragmented Source Data
FHIR Standardization Layer
Structured Feature Extraction
AI Model Training
Reliable Clinical Decision
FHIR
Inconsistent, siloed, human-coded
Resources, codes, timestamps, patient references
Machine-readable, reusable pipelines
Generalizable, accurate predictions
Workflow-level impact
Garbage In → Garbage Out
Caption-uto executive/business systems

Managers who are evaluating AI vendor proposals should ask a direct question early in any conversation: What assumptions does this model make about the structure of our data, and what happens when our data does not match those assumptions? The honest answer reveals how brittle or how adaptable the model actually is. A FHIR-native model — one designed to consume standardized FHIR resources — will degrade less catastrophically when input data varies, because the standard itself absorbs much of that variation.

6.1.2 How FHIR Improves Model Accuracy

Model accuracy is not a single number. It is a profile of performance across different patient subpopulations, clinical scenarios, and time horizons. A readmission risk model might perform well overall but poorly for patients with complex social determinants of health, because those factors were never reliably captured in the training data. A sepsis detection model might perform well in an ICU but poorly in a general medical ward, because the feature distributions differ. FHIR does not guarantee high accuracy, but it creates the conditions under which accuracy can be measured honestly and improved systematically.

The mechanism is coverage. FHIR's resource library covers clinical domains that fragmented proprietary systems often leave disconnected: medications, allergies, immunizations, procedures, observations, conditions, encounters, care plans, and more. When these domains are linked through a common patient reference and exposed through consistent APIs, an AI model can build a longitudinal, multi-dimensional view of a patient that proprietary single-system data cannot provide. That longitudinal view is the key input for the highest-value AI use cases: predicting deterioration before it becomes a crisis, identifying care gaps before they become avoidable complications, and personalizing interventions to the actual patient rather than to a statistical average.

FHIR's use of standard terminologies reinforces this. When diagnoses are consistently coded in ICD-10, procedures in CPT, labs in LOINC, and medications in RxNorm, the model can treat codes as stable semantic labels. It can learn that LOINC code 2345-7 means a glucose serum/plasma measurement, regardless of which lab system generated it, and it can link that measurement to a Condition resource coded as E11.9 and a MedicationRequest for metformin without requiring custom mapping logic. The terminological consistency is what makes AI features portable — what was learned about a feature in one dataset transfers to another.

From a manager's standpoint, this translates into a specific advantage when evaluating multi-site AI deployments. A model trained at Site A using FHIR-structured data can be validated and deployed at Site B — also on FHIR — with dramatically less re-engineering than a model trained on Site A's proprietary schema. The validation still needs to happen; patient populations differ, clinical workflows differ, and local documentation practices differ. But the structural re-engineering does not. That is a meaningful reduction in time-to-value and in deployment risk.

6.2 AI Techniques That Benefit from FHIR

Not every AI technique benefits equally from FHIR. Some approaches depend primarily on large volumes of unstructured text; others are almost entirely dependent on structured, tabular data. Understanding which techniques map to which data requirements helps a manager set realistic expectations for what FHIR enables immediately versus what requires additional investment in data sources or infrastructure.

The techniques described in this section are not theoretical. They are in active use across health systems today. In each case, FHIR's role is identifiable and concrete — not as a vague 'enabler' but as the specific mechanism that makes the technique tractable at scale.

6.2.1 Predictive Analytics

Predictive analytics — using historical data patterns to forecast future clinical events — is the most mature category of healthcare AI. Readmission risk scoring, sepsis early warning, length-of-stay prediction, deterioration alerts, and high-risk patient identification are all predictive analytics applications that health systems have been piloting for a decade or more. The persistent challenge has been moving them from pilot to production at scale.

FHIR directly addresses the scale problem. Predictive models require features — quantified representations of patient state derived from raw data. Features for a readmission model might include: number of prior admissions in 12 months, number of active medications, presence of a social work consult, whether a follow-up appointment was scheduled at discharge, and lab values at discharge. Extracting each of these features from a FHIR-compliant data layer is a matter of querying the appropriate resources — Encounter, MedicationRequest, CarePlan, Appointment, Observation — with standardized parameters. Extracting them from a proprietary EMR database requires custom SQL against a schema that changes with every upgrade and varies across every vendor.

The operational advantage is maintainability. FHIR-based feature pipelines are stable across software upgrades because the FHIR resource structure does not change with changes to the underlying EMR version. A feature-extraction query written against the FHIR API in year one remains valid in year three, even after the EMR vendor has released two major platform updates. This stability is the difference between a predictive model that requires a data engineering team to maintain full-time and one that runs reliably with periodic validation reviews.

Managers should also note that FHIR enables real-time predictive scoring, not just batch analytics. Because FHIR APIs are designed for point-in-time queries — retrieving the current patient state for Patient/12345 — they can feed models that continuously score patients as new data arrives. A sepsis model that runs every four hours on batch data is useful. One that scores continuously as vital signs, lab results, and nursing assessments are documented is significantly more useful. FHIR's event-driven subscription capabilities — the FHIR Subscription resource — make continuous scoring architecturally feasible without requiring a custom real-time data pipeline for every new model.

Diagram 6.2 – Predictive Analytics Pipeline Powered by FHIR

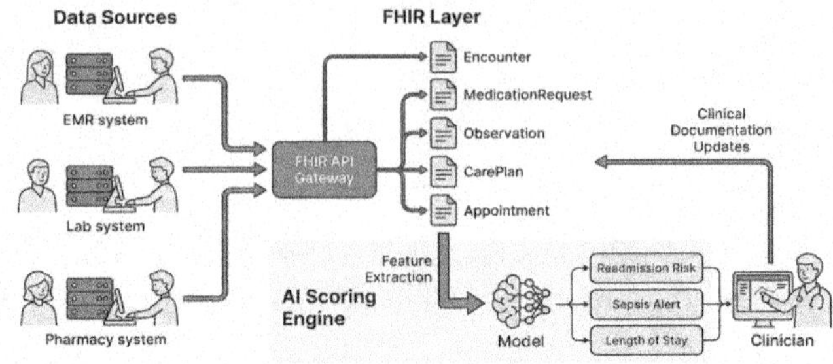

6.2.2 Natural Language Processing

Clinical notes are the richest source of information in a health record. A physician's assessment and plan, a nursing note, a radiology report, a pathology result, a social work intake — these documents contain nuance that structured data cannot capture: the patient's functional status as observed in the room, the uncertainty the clinician felt about a differential diagnosis, the family context that shaped a care decision. Natural language processing models, including large language models, are designed to extract meaning from this kind of text.

FHIR does not replace clinical text. What it does is provide the structured scaffold that makes NLP more accurate and more useful. Consider a clinical NLP application designed to extract undiagnosed conditions from physician notes. Running NLP on notes in isolation produces a list of candidate conditions — but without context, the model cannot distinguish between a condition the patient has, one the physician ruled out, or one the physician is considering as a differential. FHIR's Condition resource provides that context: if the condition already appears in the structured problem list, the NLP extraction may be redundant. If it does not, the NLP has identified a documentation gap.

The combination of FHIR-structured data and NLP also improves what engineers call 'grounding' — the ability to connect model outputs to verifiable facts about the patient. A large language model that generates a clinical summary grounded in the patient's actual FHIR resources — their current medications, their recent labs, their active diagnoses — is producing something fundamentally different from a model generating text based purely on statistical patterns in training data. The FHIR layer functions as a real-time knowledge base that tethers the model's outputs to the patient's actual clinical state.

For managers evaluating generative AI tools in clinical settings, this distinction is critical. An AI-generated clinical summary that is grounded in FHIR resources can be audited: the facts in the summary can be traced to specific resources, specific timestamps, specific authors. A summary generated without that grounding cannot. The auditability question is not just a compliance concern — it is a liability and patient safety question that boards and clinical leadership will ask, and that vendors should be able to answer specifically.

6.2.3 Machine Learning on Structured Data

The workhorse of clinical AI is not a large language model or a deep neural network. It is gradient boosting, logistic regression, and random forests — classical machine learning techniques applied to structured tabular data. These methods dominate production clinical AI for a straightforward reason: they are interpretable, they perform well on the sample sizes that health systems actually have, and they degrade predictably when input distributions shift. Clinicians and risk managers can understand them well enough to trust them in practice.

FHIR is purpose-built for this kind of machine learning. The resource model produces tabular features naturally: each patient-encounter pair becomes a row, each resource element becomes a potential column. FHIR's consistent field names, data types, and code systems mean that feature engineering can be systematized. Teams at one organization have developed FHIR-to-feature libraries — open-source tools that convert FHIR patient data into machine learning-ready dataframes — precisely because the standardization makes such generalization possible.

The mission-aligned implication for managers is that FHIR investment does not benefit only one AI project. A health system that builds a FHIR-based data layer for its readmission model is simultaneously building the infrastructure for its no-show prediction model, its medication adherence model, its surgical complication risk model, and whatever the next high-priority use case turns out to be. The infrastructure amortizes across every structured ML project the organization ever runs. That reusability changes the ROI calculation for FHIR implementation: it is not a single-project cost but a platform investment.

Managers should also be aware of the federated learning opportunity that FHIR enables. Federated learning allows multiple organizations to train a shared model without sharing patient data — each site trains on its local data and shares only model parameters. For this to work across sites, each site must produce features in a compatible format. FHIR standardization is what makes that feature compatibility achievable without a multi-year data harmonization project. Several national research networks are already using FHIR as the common data model for federated clinical AI, and the results — in terms of model performance on rare conditions that no single site could train on alone — are significant.

6.2.4 Ambient Intelligence

Ambient intelligence refers to AI systems that operate continuously in the background, monitoring conditions and surfacing relevant information without requiring explicit queries. In healthcare, this includes ambient documentation tools that listen to patient-provider conversations and generate draft notes, passive monitoring systems that analyze continuous vital sign streams for early deterioration signals, and context-aware clinical decision support that surfaces relevant information based on what the provider is doing at a given moment.

FHIR's role in ambient intelligence is architectural. Ambient systems need to know who the patient is, what their current state is, and what the clinical context demands — in real time. FHIR APIs provide exactly this: a ClinicalImpression resource that the ambient documentation tool writes draft content to, an Observation resource that receives continuous monitoring data, a Patient resource that anchors the entire encounter context. The ambient system is not working from static data or batch exports; it is reading and writing to a live FHIR layer that reflects current clinical reality.

The workflow-level impact of ambient intelligence powered by FHIR is measurable. Documentation time reduction is the most commonly cited metric — ambient documentation tools have demonstrated reductions of 30 to 50 percent in time spent on clinical note completion in published evaluations. But the deeper impact is cognitive: when the ambient system handles the documentation burden, the clinician's attention is freed for the patient. When that system is grounded in the patient's actual FHIR record — current medications, prior visit notes, active problems — the draft documentation it produces requires fewer corrections and creates less liability than documentation generated without that grounding.

Diagram 6.3 – Four AI Technique Categories and Their FHIR Dependencies

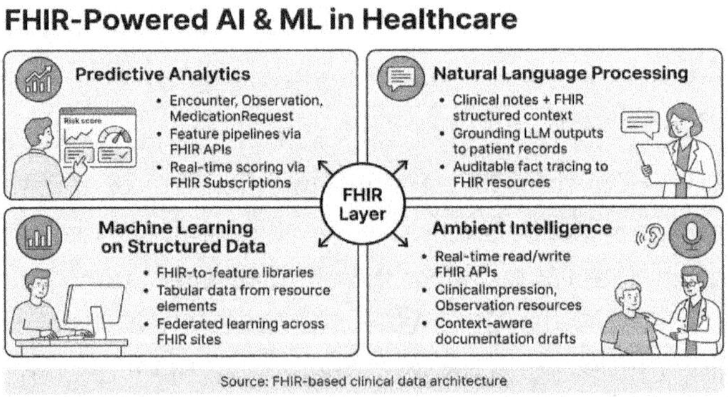

6.3 Safety and Bias Considerations

AI in healthcare carries risk that is qualitatively different from AI in most other domains. A recommendation engine that misjudges a consumer's product preferences creates a suboptimal shopping experience. An AI system that misjudges a patient's risk level can contribute to delayed treatment, inappropriate discharge, or resource allocation decisions that systematically disadvantage certain patient populations. Managers who treat AI safety as a compliance checkbox — something to hand off to legal or IT — are misunderstanding the nature of the risk.

FHIR does not make AI safe by itself. It does, however, create infrastructure that makes safety work more tractable. Responsible by design means building the auditability, traceability, and bias-detection capacity into the system from the beginning — and FHIR's structured, identifiable data is the foundation that makes all of those capabilities possible.

6.3.1 Data Representativeness

A model is only representative of the population it was trained on. This is not a theoretical concern in healthcare AI — it is a documented, recurring failure mode. Published research has found that commercial risk scores used widely across U.S. health systems systematically underestimated illness severity in Black patients relative to White patients with the same clinical profile. The root cause was training data: the models were trained on data in which healthcare utilization was used as a proxy for health need, and because Black patients have historically had less access to healthcare, the models learned to assign them lower risk scores. The data reflected systemic inequity, and the model reproduced it.

FHIR's contribution to representativeness is coverage. When data is structured and accessible through standard APIs, it becomes possible to audit the training dataset systematically. What proportion of training patients are from rural zip codes? What proportion have Medicaid versus commercial insurance? What is the racial and ethnic distribution? What languages are documented? These questions are answerable only when demographic and social data is captured in structured, queryable form — in FHIR, through the Patient resource, the Observation resource for social determinants, and the Coverage resource for insurance.

Managers should require representativeness audits as a standard deliverable in any AI procurement or development process. The audit should cover not just the overall training population but the subgroup performance of the model — does it perform equally well across race, ethnicity, age, insurance status, and geographic location? If a vendor cannot provide subgroup performance data, that is a meaningful risk signal. A FHIR-based data layer enables this data to be produced internally, without relying entirely on vendor-provided assessments.

Representativeness is also a longitudinal concern. Patient populations change. Geographies change. Clinical protocols change. A model trained on pre-pandemic data may not represent the post-pandemic patient population in terms of comorbidity burden, mental health presentation, or care utilization patterns. FHIR-based monitoring pipelines can track this drift systematically — comparing the distribution of key features in the current patient population against the training population and alerting when the gap becomes clinically significant. This kind of ongoing validation is the operational infrastructure of responsible AI deployment.

6.3.2 Bias Mitigation Strategies

Identifying bias is a necessary first step; mitigating it requires a different set of actions. Bias in healthcare AI operates at multiple levels: in the data itself (what was documented, what was coded, what was missing), in the feature selection (which variables the model is allowed to use), in the objective function (what the model is optimizing for), and in the deployment context (how the model's outputs are presented to clinicians and what actions they are expected to take). Addressing bias at only one level while ignoring the others is insufficient.

At the data level, FHIR enables several practical mitigation strategies. Missing data — one of the primary mechanisms through which historical inequity enters AI training — can be audited systematically. When social determinants of health are captured in FHIR Observation resources, it becomes possible to measure which patient segments have complete SDOH data and which do not, and to apply appropriate imputation or reweighting strategies during training. When race and ethnicity are captured in the Patient resource using standard codes, it becomes possible to perform subgroup analyses rather than treating race as a latent variable.

- Audit missing data by subgroup before training: identify which patient populations have systematically incomplete records and address through targeted data collection or imputation.
- Evaluate feature proxies: if a model uses healthcare utilization as a feature, assess whether that feature acts as a proxy for race, income, or insurance status and consider alternatives.
- Apply fairness constraints during training: modern ML frameworks support fairness-aware training objectives that can bound performance disparity across demographic groups.
- Perform subgroup validation before deployment: require the model developer to demonstrate performance parity — or explicitly acknowledge and quantify disparity — across defined patient subgroups.
- Monitor for post-deployment drift by subgroup: set up monitoring pipelines that track model performance separately for key demographic groups over time.
- Build feedback loops into the FHIR layer: when clinicians override model recommendations, capture those overrides as structured data so they can be analyzed for systematic patterns.

These strategies are not unique to FHIR, but they are more feasible with FHIR. The common thread is that they require structured, queryable data about patients' demographic characteristics, clinical history, and social circumstances — precisely the data that FHIR's Patient,

Observation, and Coverage resources are designed to capture.

6.3.3 Transparency and Explainability

Explainability in clinical AI is both a regulatory expectation and a clinical necessity. Clinicians who receive an AI recommendation without any explanation of how it was generated are in a difficult position: they cannot assess whether the recommendation is plausible given the specific patient in front of them, and they cannot exercise the kind of informed clinical judgment that patient safety requires. Regulatory frameworks — including FDA guidance on software as a medical device and the EU AI Act — are moving toward explainability requirements for high-risk AI, which encompasses most clinical decision support.

FHIR supports explainability in a specific and practical way: it provides traceable inputs. When a model's risk score is accompanied by the FHIR resources that drove the top contributing features — 'this score is elevated primarily because of three hospitalizations in the past six months (Encounter resources), an active prescription for a high-risk medication (MedicationRequest), and a hemoglobin A1c of 9.2 recorded two weeks ago (Observation)' — the clinician has something they can evaluate against their own clinical knowledge. The explanation is not an abstraction; it is a set of facts the clinician can verify in the record.

This kind of traceable explanation is only possible when the model's inputs are structured and identifiable — when 'hemoglobin A1c of 9.2' is not a pattern in unstructured text but a specific Observation resource with an ID, a timestamp, and a LOINC code. FHIR provides that identifiability. The explanation layer that surfaces the top features can link directly to the underlying FHIR resources, creating an audit trail from clinical decision back to data source that satisfies both clinical and regulatory standards.

For the manager, explainability has a practical governance implication. When a model makes a recommendation that contributes to an adverse patient outcome, the post-event review will ask: what did the model know, when did it know it, and why did it recommend what it recommended? That question is answerable — or not — based on whether the model's inputs and reasoning are traceable. FHIR-grounded AI is traceable. AI built on unstructured or proprietary data is, at best, partially traceable. Governance committees, risk officers, and patient safety teams should treat traceability as a non-negotiable architectural requirement.

Diagram 6.4 – Bias and Safety Governance Framework for FHIR-Powered AI

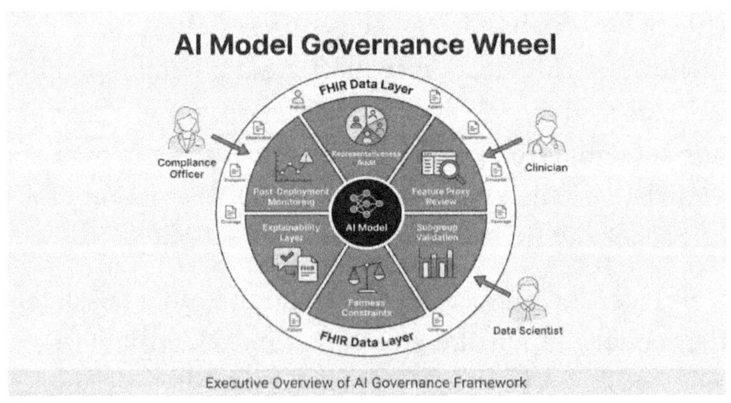

Executive Overview of AI Governance Framework

6.4 FHIR as AI Fuel

The metaphor of fuel is deliberate. AI models are engines: they have the capacity to process inputs and produce outputs at scale, but they need the right kind of fuel to run. Unstructured, fragmented, inconsistently coded data is low-octane fuel — the engine runs, but it runs inefficiently, it produces poor outputs, and it requires constant maintenance to keep it functioning at all. FHIR-standardized data is high-octane fuel: the engine runs more efficiently, produces better outputs, and scales without requiring proportional increases in maintenance effort.

This section examines what FHIR as AI fuel means in practice — how it accelerates the training process, what real-world implementations look like, and why standardization is the precondition for AI that scales beyond a single institution or a single use case.

6.4.1 How Standardized Data Accelerates Training

Training a machine learning model begins with data acquisition and preparation, not with model design. For most healthcare AI projects, this phase is the longest. Data must be identified, accessed, extracted, cleaned, normalized, and structured before a single line of modeling code is written. The time this takes varies enormously depending on the state of the organization's data infrastructure — but in organizations without FHIR, it commonly takes six to twelve months just to get a training dataset ready. In organizations with mature FHIR APIs, the same process can take weeks.

The acceleration comes from several compounding factors. First, FHIR APIs provide direct, authenticated access to structured data without requiring database administrators to extract and export custom datasets. An AI engineer with appropriate authorization can query the FHIR server for all Encounter resources matching specific criteria — admission date ranges, encounter types, diagnosis codes — and receive structured JSON responses that are immediately usable. No export request, no ticket to the data warehouse team, no weeks-long wait for an IT project.

Second, FHIR's standard resource definitions mean that feature extraction code does not need to be written from scratch for each new project. A library of FHIR-to-feature transformations — functions that convert Observation resources into numerical features, that count Encounter resources by type and time window, that extract active MedicationRequest entries for polypharmacy scoring — can be built once and reused across every project. Each new AI initiative starts from a library of existing, validated transformations rather than from zero.

Third, FHIR's patient-centric data model simplifies the most time-consuming aspect of data preparation: record linkage. In fragmented data environments, linking records for the same patient across multiple systems — matching on name, date of birth, medical record number, and other identifiers — can consume a substantial fraction of the data preparation budget. When implemented consistently, FHIR's Patient resource provides a master patient identity that links all other resources. The linkage problem is solved at the infrastructure layer rather than in every individual AI project.

1. Define the clinical question and identify the FHIR resources that capture the relevant data elements — for example, Observation for labs, Condition for diagnoses, Encounter for utilization.
2. Authenticate to the FHIR server using SMART on FHIR OAuth scopes appropriate to the data access level required for training.
3. Execute FHIR API queries to retrieve the training cohort, applying date range, diagnosis code, and demographic filters as search parameters.
4. Apply standardized FHIR-to-feature transformation libraries to convert resource JSON into machine learning-ready tabular data.
5. Perform representativeness audit on the resulting dataset before training begins — check subgroup distributions, missing data rates, and code coverage.
6. Train, validate, and document the model with full traceability back to the FHIR resource queries that produced the training data.

The sequence above is not a theoretical ideal — it is the architecture that organizations with mature FHIR infrastructure are already executing. The difference between executing this sequence in six weeks versus six months is almost entirely a function of FHIR readiness. Managers who have made an FHIR investment are making an AI investment, whether or not they have yet articulated it that way.

6.4.2 Real-World Examples of AI Powered by FHIR

The evidence base for FHIR-powered AI is no longer limited to research publications and proof-of-concept demonstrations. Health systems, payers, life sciences organizations, and digital health companies are building and deploying AI systems on FHIR infrastructure at scale. The examples below illustrate the range of applications and the common role that FHIR standardization plays in each.

The National COVID Cohort Collaborative (N3C) aggregated FHIR-based patient data from over 75 U.S. clinical sites to build one of the largest COVID-19 research datasets in the world. The use of FHIR as the common data model was central to the project's ability to harmonize data across sites with different EMR vendors and data structures. Within that dataset, AI models were trained for mortality prediction, long COVID identification, and treatment response analysis — research that would have been infeasible if each site's data had remained in proprietary formats.

Several major health systems have deployed FHIR-based ambient documentation tools that use AI to generate draft clinical notes from physician-patient encounter audio. The FHIR layer provides the patient context — current medications, active diagnoses, prior notes — that allows the ambient AI to generate contextually accurate drafts rather than generic templates. The structured output is then written back to the EMR via FHIR APIs, creating a complete audit trail from audio input to signed note.

In the payer space, AI-driven prior authorization platforms are using FHIR APIs to pull structured clinical data — diagnoses, procedures, lab results, medications — directly from provider EHR systems, replacing the manual chart abstraction process that has historically delayed authorization decisions by days. The AI assesses clinical necessity against policy criteria using structured FHIR data rather than scanned PDFs, reducing processing time from days to hours and improving accuracy by eliminating transcription errors.

Life sciences companies are using FHIR-structured real-world data to accelerate clinical trial site selection, patient recruitment, and safety signal detection. When trial eligibility criteria are translated into FHIR search queries — specific diagnosis codes, medication histories, lab value thresholds — sites can identify eligible patients in days rather than weeks, without requiring manual chart review. Post-market safety surveillance that would previously have required manual case processing can be partially automated through FHIR-based adverse event monitoring pipelines.

Population health management platforms are using FHIR to aggregate data from multiple EHR systems, claims databases, and social determinants sources into unified patient profiles that feed AI-driven care gap identification, outreach prioritization, and risk stratification. The ability to query across systems using a common API surface — rather than building system-specific integrations for each data source — is what makes multi-source population health analytics operationally feasible.

Diagram 6.5 – Real-World FHIR-Powered AI Use Cases Across the Healthcare Ecosystem

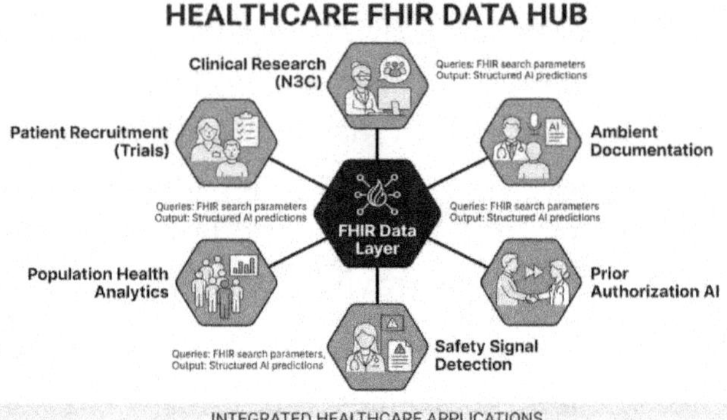

6.4.3 Why FHIR Unlocks Scalable Intelligence

Scale is where most healthcare AI initiatives fail. A model performs well at a single site, with a single EMR vendor, in a carefully controlled pilot. Then the organization tries to deploy it across a second site, or across a health system network, or to sell it to external customers — and the model requires extensive re-engineering for every new environment. Each new site has different data formats, different coding practices, different missing data patterns. The model that worked in Pittsburgh does not work in Phoenix without months of additional work.

FHIR is the solution to this scaling problem. When every site in a network exposes patient data through FHIR APIs — with consistent resource types, standard terminologies, and validated implementation guides — the AI infrastructure that works at one site transfers to every other site without fundamental re-engineering. The feature extraction pipeline is the same. The data validation checks are the same. The monitoring infrastructure is the same. What varies is the patient population, and that variation should be addressed through site-specific model validation, not through site-specific data engineering.

The scalability also extends to the AI model ecosystem. Because FHIR provides a common data interface, AI vendors can build products that target FHIR as their input layer rather than integrating directly with individual EMR systems. This shifts the integration burden from AI vendors to health systems — but it shifts it to a standardized integration that health systems only need to perform once. A health system that has implemented FHIR can onboard a new AI vendor in days rather than months, because the vendor does not need a custom integration; they need FHIR credentials. This dynamic is already reshaping the healthcare AI marketplace, with FHIR-native AI tools increasingly crowding out proprietary-integration-dependent alternatives.

For network operators — integrated delivery networks, ACOs, large multi-hospital systems — the scalability of FHIR-powered AI has direct financial implications. A model deployed across thirty hospitals on a common FHIR platform generates thirty times the data for ongoing training and validation. It also generates thirty times the improvement signal: when clinicians across the network override or accept model recommendations, those signals flow back through the FHIR layer as structured data that can be used to retrain the model on real-world performance rather than on historical training data alone. The network effect of FHIR-standardized AI is not linear — it is compounding.

Managers building enterprise AI strategies should view FHIR standardization as a prerequisite for compound returns. The first AI model deployed on a FHIR platform delivers value on its own terms. The second delivers additional value but costs less to build. The tenth delivers the most value of all because it benefits from the full maturity of the feature library, the monitoring infrastructure, the deployment pipeline, and the institutional knowledge about how to validate and govern AI in this specific clinical environment. That trajectory — from expensive first mover to efficient serial deployer — is what the pilot-to-production pathway on FHIR looks like.

Diagram 6.6 – The Compounding Value of FHIR-Standardized AI Across a Network

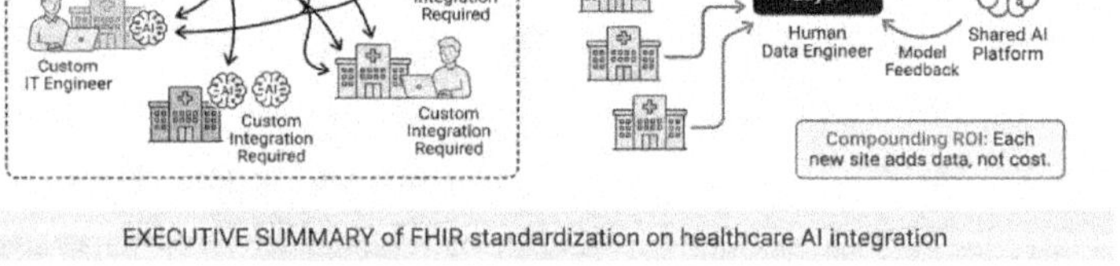

EXECUTIVE SUMMARY of FHIR standardization on healthcare AI integration

6.5 Takeaway

Healthcare AI is not a technology problem with a technology solution. It is a data problem with a data solution

— and FHIR is that solution. The single most important thing a health system can do to improve the quality, safety, and scalability of its AI initiatives is to invest in FHIR infrastructure: implementing FHIR-compliant APIs across its major platforms, enforcing standard terminologies, capturing structured data for clinical domains that were previously documented only in narrative text, and building FHIR-based data pipelines that AI engineers can actually work with.

This is not a call to delay AI until FHIR infrastructure is perfect. Perfection is not achievable, and waiting for it is itself a decision — one that lets competitors and peer institutions move ahead. The practical path is to pursue FHIR implementation and AI development in parallel, letting each inform the other. AI projects will surface data gaps that FHIR implementation should address. FHIR implementation will open AI opportunities that were previously inaccessible. The two initiatives are not sequential; they are symbiotic.

For managers making decisions right now, the chapter's central argument comes down to six realities:

- Structured data reduces AI project cost by compressing the data preparation phase — the phase that consumes 60 to 80 percent of most AI project budgets.
- FHIR's standard resource types — Observation, Condition, Encounter, MedicationRequest, Patient — create reusable feature extraction infrastructure that amortizes across every AI project the organization runs.

- Bias and safety risks in clinical AI are data problems: they originate in missing, inconsistent, or unrepresentative training data and are mitigated by the structured, auditable data that FHIR provides.
- Explainability and traceability — which regulators, clinicians, and patients are increasingly demanding — require that model inputs be structured and identifiable, not buried in unstructured text or proprietary schemas.
- Scalability across sites and use cases is achievable on FHIR in a way it is not on proprietary or siloed data, because FHIR eliminates per-site data engineering as the bottleneck for AI deployment.
- The network effect of FHIR-powered AI compounds: each additional site, each additional model, and each additional data source increases the value of the platform without proportionally increasing the cost.

These realities have a management implication that is both strategic and immediate. Strategically: FHIR investment is AI investment, and organizations that have not yet made that connection are undervaluing their interoperability work and underestimating their AI readiness. Immediately: every AI initiative currently under evaluation should be assessed against the FHIR readiness of the underlying data. A model that requires custom data integration is a model that will cost more to deploy, more to maintain, and more to scale than its proposal document suggests.

The question for the manager is not whether AI will become central to healthcare operations — it already is. The question is whether the data infrastructure exists to support AI that is accurate, safe, explainable, and scalable. FHIR is the answer to that question. The organizations that understand this — and act on it — will be the ones that move from perpetual pilot to durable, mission-aligned production AI. Everything else is preparation.

AI systems learn from data. FHIR determines the quality, completeness, and structure of that data. Investing in FHIR is investing in every AI model your organization will ever build.

7 Real-World Use Cases: Where FHIR + AI Are Already Working

Every chapter in this book has built toward a single question: what does the FHIR and AI pairing actually look like in production? Not in a whitepaper, not in a vendor demo, but in the day-to-day workflows of a health system trying to keep patients safe, control costs, and comply with regulations. This chapter answers that question directly. It walks through the use cases where FHIR-enabled AI is already demonstrating workflow-level impact across four domains: clinical care, administrative operations, patient-facing experiences, and population health.

None of these use cases require magic. They require structured, accessible, timely data — exactly what FHIR provides — and they require AI models that are purpose-built, validated, and integrated at the point of care or decision. For managers, the practical question is not whether these systems work in theory; the evidence is accumulating. The question is how to evaluate readiness, how to structure the pilot-to-production pathway, and how to assign accountability when the system is live.

This chapter is organized to support that evaluation. Each domain section covers the core workflows, the data dependencies, the failure modes that managers must anticipate, and the organizational conditions that separate successful deployments from expensive mistakes. Wherever possible, the examples are grounded in real enterprise or public-sector settings. The goal is operational clarity: by the end of this chapter, you should be able to match a use case to your organization's current capability and make an informed decision about sequencing.

7.1 Clinical Use Cases

Clinical use cases carry the highest stakes in any AI deployment program. A decision support alert that fires at the wrong moment can be ignored, contributing to alert fatigue. A risk score that misclassifies a patient can delay intervention. A medication recommendation built on incomplete data can cause harm. Healthcare leaders who treat clinical AI as just another software rollout are setting themselves up for both clinical and regulatory exposure. The appropriate lens is clinical governance first, technology second.

That said, clinical AI is also where FHIR's benefits are most visible and most immediate. When a health system has a FHIR-compliant data layer — with structured clinical data flowing from the EHR, the pharmacy, the lab, and the monitoring system in near real time — an AI model has the raw material it needs to generate actionable, accurate output. The organizations that have advanced furthest in clinical AI are almost universally those that invested earliest in their data infrastructure, not their models.

7.1.1 Clinical Decision Support

Clinical decision support (CDS) is the oldest form of clinical AI, and FHIR is now reshaping how it is built and deployed. Traditional CDS was rule-based: if a patient has a penicillin allergy flag and a prescription for amoxicillin is entered, generate an alert. That model works well for deterministic rules but cannot handle the complexity of real clinical scenarios — comorbidity combinations, drug-drug-disease interactions, timing of interventions, or population-specific risk thresholds.

Modern CDS leverages machine learning models trained on longitudinal patient data to generate probabilistic recommendations. These models require access to rich, structured clinical data: current medications from the MedicationRequest resource, active conditions from the Condition resource, recent lab values from the Observation resource, vital sign trends from the Observation resource, and prior encounters from the Encounter resource. FHIR provides a uniform way to query and compose this information, regardless of which EHR vendor is in use.

The CDS Hooks specification, which sits on top of FHIR, enables real-time, in-workflow integration. A CDS Hook fires at a defined point in the clinical workflow — when a medication is prescribed, when a patient is admitted, when a care plan is updated — and the AI service responds with a card that surfaces a recommendation or flag directly in the clinician's interface. No tab switching, no separate application, no secondary login. The alert appears in context, at the moment it matters.

For managers evaluating CDS programs, the critical governance questions are: Who approved the model? What was the validation cohort? How often is the model retrained and against what benchmark? Who receives notification when the alert rate drops or spikes unexpectedly? These are not IT questions — they are clinical governance questions, and the answers must exist before the system goes live.

- Sepsis early warning: AI models trained on vital signs, lactate trends, WBC counts, and nursing assessments can flag deteriorating patients 6–12 hours before clinical criteria are met. The FHIR Observation and Condition resources feed the model in real time. Health systems using these tools report 20–30% reductions in sepsis mortality in published pilots.

- Drug-drug interaction (DDI) intelligence: Beyond simple rule-based DDI flags, AI systems can weigh interaction severity against patient-specific factors — renal function, body weight, current lab values — and generate context-aware recommendations rather than generic warnings.

- Diagnostic support: Radiology AI reading chest X-rays, pathology AI scanning slides, and NLP systems extracting findings from unstructured notes are all downstream consumers of FHIR-structured clinical context. An AI chest X-ray that can access the patient's current medications, smoking history, and prior imaging findings generates fewer false positives than one that works from the image alone.

- Care gap alerts: CDS systems can surface unmet preventive care needs at the point of contact — a patient due for a colorectal cancer screening who presents for a hypertension follow-up. The alert is mission-aligned with value-based care targets and requires only scheduling workflow integration, not a complex clinical model.

Diagram 7.1 – Clinical Decision Support: FHIR-Enabled CDS Hooks Workflow

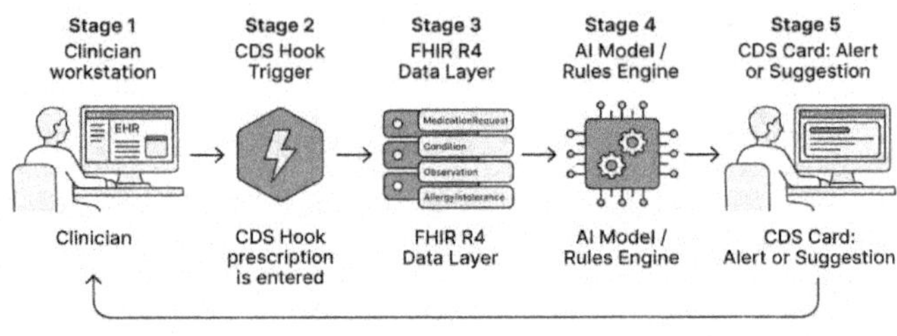

Seamless integration of CDS into the clinical workflow optimizes decision-making.

7.1.2 Care Coordination

Care coordination fails when information doesn't travel with the patient. A patient discharged from a hospital who sees a primary care physician three days later relies on a fax, a summary PDF, or the patient's own memory to communicate what happened. The primary care physician makes decisions without complete information. This is not a rare failure mode — it is the dominant mode of care transition in most health systems today.

FHIR-based care coordination changes this by making the patient's record a living, queryable resource rather than a static document. When a patient is discharged, FHIR resources — Encounter, Procedure, MedicationRequest, CarePlan, and Observation — can be pushed to a care coordination platform in near real time. AI systems can then analyze these resources to identify high-risk transitions, generate automated follow-up tasks, and alert care managers who need to intervene before the patient returns to the emergency department.

The practical value of AI in care coordination is in prioritization. A care management team serving 5,000 attributed patients cannot manually review every discharge. An AI model that ranks patients by 30-day readmission risk — using FHIR data on prior admissions, medication adherence signals, social determinants, and comorbidity burden — allows care managers to focus their limited capacity on the patients who need them most. This is not automation for automation's sake; it is a workforce multiplier that supports mission-aligned outcomes.

Large health systems using FHIR-connected care coordination platforms report measurable reductions in 30-day readmission rates, particularly for high-risk populations such as heart failure, COPD, and post-surgical patients. The workflow-level impact is tangible: care managers receive daily work lists sorted by AI-generated priority scores, with drill-down access to the FHIR data that drove the score. This transparency is not just operationally useful — it is a prerequisite for clinical staff to trust and act on the recommendations.

Managers should note that care coordination AI is heavily dependent on the completeness of the underlying FHIR data. A health system that has only partially implemented FHIR or that has significant data quality gaps in its Encounter or MedicationRequest resources will generate unreliable risk scores. Data completeness audits should precede any care coordination AI deployment — not follow it.

7.1.3 Risk Stratification

Risk stratification is the process of segmenting a patient population by predicted health risk to allocate care resources appropriately. It is foundational to value-based care programs, care management programs, and population health management. Historically, risk stratification relied on claims-based algorithms — simple models that used diagnosis codes and utilization patterns to estimate future cost. These models are slow (claims data lags by weeks to months), incomplete (they miss clinical detail), and retrospective by design.

FHIR-enabled AI risk stratification operates on clinical data in near real time. A risk model with access to current lab values, medication lists, vital sign trends, and recent encounter notes can generate a more accurate and more timely risk score than any claims-based algorithm. When combined with social determinants data — housing stability, food security, transportation access — the model can identify patients whose clinical risk is compounded by social vulnerability, enabling more targeted interventions.

The operational implications are significant. A health system running an accountable care organization (ACO) needs to know which patients are approaching the cost and utilization thresholds that will affect shared savings calculations. A Medicaid managed care organization needs to identify members who are likely to generate high emergency utilization. A specialty practice operating under a bundled payment arrangement needs to identify post-procedure patients at highest risk for complications before they manifest. All of these programs benefit from risk stratification that is faster, richer, and more actionable than claims-based alternatives.

For managers, the equity dimension of risk stratification is not optional. AI risk models trained on historical claims or clinical data can encode and amplify existing disparities if not carefully designed and validated. A model that systematically under-scores risk for underserved populations — because they have lower historical utilization, not lower need — will direct resources away from those populations. Responsible by design means requiring bias audits as part of the model evaluation process, segmenting performance metrics by race, ethnicity, and socioeconomic status, and establishing a governance process to act on findings.

7.1.4 Medication Management

Medication errors remain one of the leading causes of preventable patient harm. The Joint Commission and Institute for Safe Medication Practices have documented for decades that the complexity of modern pharmacotherapy — polypharmacy in elderly patients, weight-based dosing in pediatrics, renally-adjusted dosing in kidney disease, narrow therapeutic index drugs requiring monitoring — exceeds what any individual clinician can reliably manage from memory alone.

FHIR provides the data substrate for AI-driven medication management by making the medication list, allergy list, current lab values, and relevant conditions available as structured, queryable resources. An AI medication management system can synthesize this data to support several high-value functions: detecting potential adverse drug events before they occur, identifying medication adherence gaps from fill records and patient-reported outcomes, optimizing dosing for high-risk drugs based on current renal and hepatic function, and flagging therapeutic duplications across the reconciled medication list.

Medication reconciliation — the process of comparing the patient's medication list against the admission orders, discharge orders, or care transition documents — is a particularly high-value AI application. Automated reconciliation systems that use FHIR MedicationRequest, MedicationStatement, and MedicationDispense resources can identify discrepancies that human reconciliation misses, especially in high-volume, high-complexity settings like transitions from hospital to skilled nursing facility.

Pharmacy-integrated AI systems are now in production at major health systems, generating real-time pharmacist alerts for high-risk medication combinations, supporting stewardship programs for antibiotics and opioids, and providing clinical decision support at the point of prescribing. The pilot-to-production pathway for medication AI tends to move faster than other clinical AI because the value proposition is well-understood, the liability framework is established, and pharmacists are experienced at working with clinical decision support tools.

- Opioid stewardship: AI models that identify patients at high risk for opioid misuse using prescription history, behavioral health diagnoses, and social determinants, triggering clinical review before a prescription is filled.

- Antibiotic stewardship: AI systems that recommend antibiotic de-escalation based on culture results flowing through the FHIR Observation resource, reducing broad-spectrum antibiotic overuse and its downstream resistance risks.

- Polypharmacy management: In patients taking ten or more medications, AI-assisted deprescribing tools flag medications that may be no longer indicated, have significant interaction risk, or are not supported by the patient's current clinical status.

- Adherence prediction: Using pharmacy dispense records, patient portal engagement data, and social determinants, AI models can predict non-adherence before it results in clinical deterioration, enabling proactive outreach.

Diagram 7.2 – Medication Management: AI-Driven Reconciliation and Safety Loop

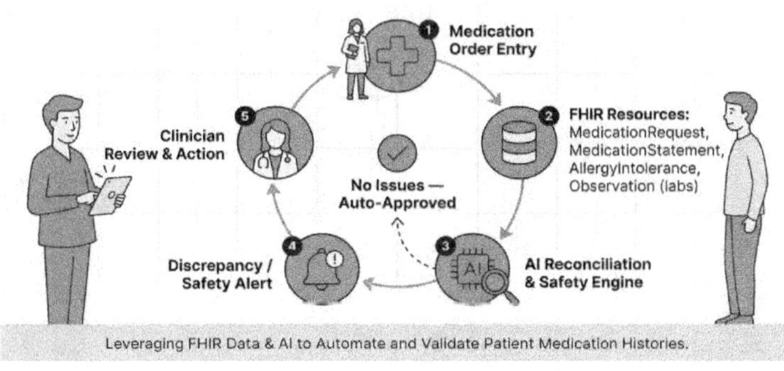

7.2 Administrative Use Cases

Administrative functions consume a disproportionate share of healthcare spending. The United States spends an estimated $265 billion annually on healthcare administrative costs — billing, prior authorization, credentialing, claims management, and compliance — that produce no clinical value. FHIR and AI together create a credible path to substantially reducing that burden, not by replacing human judgment, but by automating deterministic work, accelerating ambiguous work, and eliminating re-keying, faxing, and manual lookups that currently consume staff time.

Administrative AI is often the fastest path to demonstrable return on investment in a FHIR deployment program. The use cases are well-defined, the data flows are quantifiable, and the financial impact is measurable in weeks rather than years. For managers building a business case for FHIR infrastructure investment, administrative use cases frequently provide the near-term ROI that justifies the longer-term clinical AI roadmap.

7.2.1 Prior Authorization Automation

Prior authorization is widely regarded as the most operationally painful intersection between health systems and payers. A physician orders a procedure, a medication, or a specialist referral. The health system's staff must then gather clinical documentation, submit it to the payer in a specific format, wait for a response that may come back as a request for more information, re-submit, and wait again. The average prior authorization takes 1.9 business days, occupies significant physician and administrative staff time, and frequently delays care in ways that affect patient outcomes.

The FHIR-based prior authorization workflow changes this structure fundamentally. The Da Vinci Project, a collaboration between health systems, payers, and vendors, has produced FHIR-based implementation guides specifically for prior authorization — the Coverage Requirements Discovery (CRD) and Documentation Templates and Rules (DTR) guides. These allow a health system's EHR to query a payer's FHIR server in real time to determine whether a proposed service requires prior authorization, what documentation is required, and whether the documentation already in the patient's chart satisfies those requirements, all before the clinician completes the order.

AI layers on top of this FHIR workflow at several points. Natural language processing can extract relevant clinical evidence from unstructured notes to populate the required documentation automatically. Machine learning models trained on approval and denial patterns can predict authorization likelihood, flag cases that require human escalation, and recommend documentation strategies for borderline cases. Robotic process automation handles the structured data submission to payers who have not yet adopted FHIR-based APIs.

Health systems that have implemented FHIR-enabled prior authorization automation report significant reductions in administrative burden: authorization processing times dropping from days to hours for many service categories, denial rates falling as documentation quality improves, and staff reallocated from data entry to exception management. The workflow-level impact is not incremental — it restructures the authorization team's work from manual processing to supervisory oversight.

1. Map current authorization volume by service category and payer to identify where automation will have the greatest ROI.
2. Confirm payer FHIR API availability using the CMS Interoperability Rule payer directory and evaluate Da Vinci CRD/DTR readiness for your top three payers.
3. Assess EHR vendor's native support for CDS Hooks and FHIR-based prior authorization workflows, or identify middleware integration options.
4. Pilot automation for a single high-volume, high-denial service line — imaging or specialty medications are typical starting points — with clear baseline metrics established before go-live.
5. Define escalation protocols for cases where the AI model's documentation prediction is below a confidence threshold, ensuring human review remains in the workflow for edge cases.

7.2.2 Revenue Cycle Optimization

Revenue cycle management encompasses the full financial workflow from patient registration through final payment: scheduling, eligibility verification, patient financial counseling, charge capture, coding, billing, claim submission, payment posting, and denial management. Each step in this workflow is a potential failure point — a missed charge, an incorrect code, a late filing, a payer-specific billing requirement violated — and the cumulative effect of those failures is significant. Industry estimates suggest that 10–15% of healthcare revenue is lost to billing inefficiencies and avoidable denials.

FHIR provides a structured data layer that can feed AI models at multiple points in the revenue cycle. The Claim and ExplanationOfBenefit FHIR resources represent the billing layer; the Condition, Procedure, and Encounter resources represent the clinical layer. AI systems that can read both layers simultaneously are better positioned to identify coding gaps, detect charge capture omissions, and predict claim outcomes before submission.

Computer-assisted coding (CAC) is one of the most mature applications of AI in the revenue cycle. Modern CAC systems use natural language processing to read clinical documentation and suggest ICD-10, CPT, and HCPCS codes with supporting evidence. When integrated with FHIR-structured clinical data, CAC systems can cross-validate suggested codes against the documented diagnoses and procedures, reducing the error rate and the downstream claim denial rate. Health systems that have deployed CAC broadly report coding productivity improvements of 25–40% and measurable improvements in first-pass claim resolution rates.

Denial prevention AI is an emerging priority for revenue cycle leaders. Rather than managing denials after they occur, these systems analyze claim data prior to submission to identify characteristics associated with denial — missing modifiers, unsupported medical necessity, documentation gaps, coordination of benefits conflicts — and flag them for correction before the claim is filed. The financial impact is asymmetric: preventing a denial costs far less than working it after the fact, both in staff time and in revenue timing.

7.2.3 Claims Processing

Claims processing is the engine room of both provider billing operations and payer administrative operations. For providers, it is about ensuring that submitted claims are accurate, complete, and filed on time. For payers, it is about adjudicating claims accurately, detecting fraud, waste, and abuse, and processing payments efficiently. Both sides of the transaction are under pressure to reduce costs, reduce error rates, and accelerate cycle times.

FHIR-based claims exchange, supported by the FHIR ExplanationOfBenefit and Claim resources, is enabling AI-driven automation at scale. On the provider side, AI systems can validate claims against payer-specific rules before submission, predict adjudication outcomes, and automatically resubmit corrected claims without manual intervention. On the payer side, AI systems can process high-confidence, low-complexity claims automatically — straight-through processing — while routing complex or flagged claims to human reviewers.

Fraud, waste, and abuse (FWA) detection is a major application of AI in claims processing. Machine learning models trained on claims patterns can identify anomalies — unusual billing frequencies, geographic outliers, procedure combinations that are clinically inconsistent — with greater accuracy and at greater scale than rule-based systems. When these models are fed structured FHIR data rather than flat claims files, they have access to richer context: the clinical conditions documented at the time of service, the medications prescribed, the provider's documented specialty and credentials.

For managers, claims processing AI is primarily a cost and accuracy story. The KPIs to track are: first-pass claim resolution rate, cost per claim adjudicated, denial rate by claim category, and time to payment. AI deployments in claims processing typically require strong collaboration between IT, revenue cycle operations, and compliance — the last of which is important because automated claims decisions carry regulatory and audit exposure that must be managed deliberately.

7.2.4 Eligibility and Benefits Verification

Eligibility and benefits verification is one of the highest-volume, lowest-value-added administrative tasks in healthcare. A patient scheduling an appointment requires a verification of their current insurance coverage, plan type, deductible status, copay obligations, and in-network provider status. In a large health system, this process is performed thousands of times per day, consuming significant staff time and generating errors that, when undetected, result in claim denials or unexpected patient balances weeks later.

FHIR Coverage resources and the emerging Patient Access APIs required under the CMS Interoperability Rule are creating a path toward real-time, automated eligibility verification. When a patient's payer can respond to a FHIR query at the time of scheduling with accurate coverage details, the manual verification process becomes largely unnecessary for straightforward cases. AI handles the exceptions: patients with multiple coverage sources, patients whose plan details are inconsistent with prior records, patients whose deductible status has changed mid-year.

Beyond simple eligibility, AI systems can use FHIR Coverage and ExplanationOfBenefit data to generate patient-specific financial estimates before service. A patient facing a significant out-of-pocket expense for an elective procedure deserves an accurate estimate, not a broad range that varies by thousands of dollars. AI-driven cost estimation tools that incorporate real payer contract terms, the patient's actual deductible and out-of-pocket maximum status, and procedure-specific cost data are becoming standard at health systems that compete on patient experience.

Diagram 7.3 – Administrative Automation: FHIR-Powered Prior Authorization and Eligibility Workflow

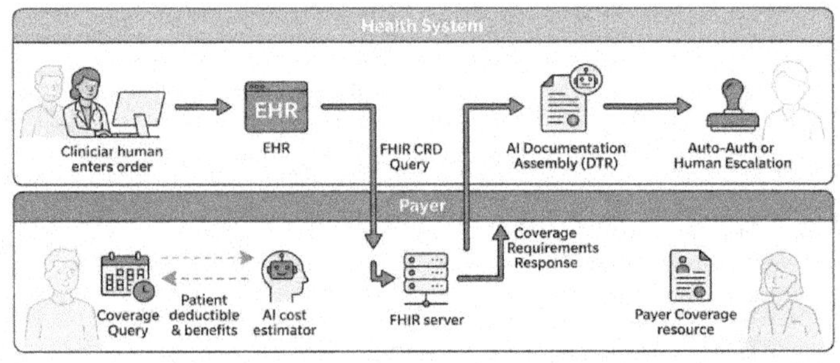

7.3 Patient-Facing Use Cases

Patients interact with healthcare organizations through a growing number of digital touchpoints: patient portals, mobile apps, wearable devices, telehealth platforms, automated messaging, and self-service scheduling systems. Each of these interactions generates data, and each represents an opportunity to deliver personalized, intelligent support. But the patient-facing domain is also where the risks of poorly designed AI are most consequential — patients acting on inaccurate health information, algorithmic personalization that excludes vulnerable populations, and digital tools that widen the access gap rather than close it.

FHIR provides the data foundation that makes patient-facing AI both possible and trustworthy. When a patient-facing application can access the patient's actual health record — through SMART on FHIR authorization and standard FHIR APIs — it can deliver experiences that are grounded in clinical reality rather than generic health content. The SMART on FHIR framework ensures that data access is patient-consented, application-scoped, and auditable. This combination of clinical relevance and privacy protection is what separates responsible patient-facing AI from the health app ecosystem's many failures.

7.3.1 Personalized Engagement

Personalized patient engagement means delivering the right message, to the right patient, at the right time, through the channel they actually use. It sounds straightforward, but most health systems still operate engagement programs that are essentially mass campaigns: the same reminder message goes to every patient regardless of their health status, communication preferences, literacy level, or prior response behavior. The result is low engagement rates and high opt-out rates.

AI-driven personalization changes this by treating engagement as a prediction problem. For each patient, the system predicts which communication channel (SMS, email, phone, app notification) is most likely to produce a response, what message content and tone will resonate, and what action is most important to prompt — a care gap closure, a medication refill, a follow-up appointment, a preventive screening. These predictions are trained on historical engagement data and refined continuously as new responses come in.

The clinical data layer that powers these predictions comes from FHIR. The AI system needs to know the patient's current health status, recent encounters, upcoming care gaps, active chronic conditions, and medication regimen. FHIR provides this through the Patient, Condition, Encounter, MedicationRequest, and Appointment resources. When a patient interaction system is authorized via SMART on FHIR, it can access exactly the data it needs without requiring a custom EHR integration for every health system it serves.

For managers, personalized engagement AI should be evaluated against three criteria: clinical impact (are the actions prompted clinically valuable?), equity (does the system perform equally across demographic groups, or does it systematically underengage certain populations?), and trust (do patients understand they are interacting with an AI-driven system, and do they have meaningful opt-out options?). A responsible by design patient engagement program addresses all three before go-live.

7.3.2 Digital Front Door

The 'digital front door' concept refers to the patient's first points of contact with a health system outside of a clinical encounter — scheduling, symptom checking, care navigation, financial counseling, and pre-registration. Historically, these entry points were either manual (call center, front desk) or minimally automated (static appointment portals with limited functionality). AI is transforming the digital front door into an intelligent, adaptive system that routes patients to the right care setting efficiently.

AI-driven symptom checkers and care navigation tools use clinical decision logic to assess symptom severity, match patients to appropriate care settings (primary care, urgent care, emergency department, telehealth, or self-care), and generate appointment requests or clinical alerts. When integrated with FHIR, these systems can personalize the triage logic: a patient with a known cardiac history who reports chest tightness gets a different recommendation than a healthy 25-year-old with the same symptom. The FHIR Condition, AllergyIntolerance, and MedicationRequest resources provide the clinical context that makes this differentiation possible.

Pre-visit AI workflows are another high-value component of the digital front door. Patients who complete structured intake questionnaires before their appointment — using tools that are pre-populated with their known health information from FHIR — arrive with better-prepared clinical encounters. The clinician enters the room with a structured summary of the patient's recent history, active concerns, and documentation gaps, rather than spending the first ten minutes collecting information that should already be in the record.

For managers, the digital front door is a patient experience investment with revenue cycle implications. Reducing no-show rates, filling schedule gaps efficiently, and routing patients to the right care setting on the first attempt all affect both patient satisfaction scores and operational efficiency. AI that integrates with FHIR Appointment, Schedule, and Slot resources can automate much of the scheduling optimization logic that currently requires staff intervention.

7.3.3 Remote Monitoring

Remote patient monitoring (RPM) generates a continuous stream of physiological data — blood pressure, glucose, oxygen saturation, heart rate, weight, activity, sleep — from devices that patients carry or wear in their daily lives. Without an intelligent data management layer, this stream is overwhelming: clinicians cannot manually review thousands of data points per patient per week. AI is the only practical way to convert RPM data into actionable clinical information at scale.

FHIR Observation resources are the natural container for RPM data. Device-generated readings can be mapped to FHIR Observation instances with standard LOINC codes — blood pressure, blood glucose, pulse oximetry — and stored in a FHIR server where AI models can process them continuously. When a patient's blood pressure trend crosses a clinically significant threshold, or when a heart failure patient's weight increases by more than two pounds in a day, the AI model generates a clinical alert routed to the appropriate care team member.

The clinical use cases for RPM with AI are concentrated in chronic disease management: heart failure, diabetes, COPD, hypertension, and post-surgical recovery. These conditions share a common pattern — they are stable most of the time, but deterioration follows predictable signals that, if detected early, allow intervention before hospitalization becomes necessary. FHIR-structured RPM data, processed by AI, is the mechanism for making that early detection scalable.

The reimbursement environment has improved significantly for RPM programs. CMS has established billing codes for remote physiologic monitoring services, providing a financial foundation for building sustainable programs. For managers, the business case for RPM AI rests on three legs: readmission reduction (which has direct financial and quality implications under value-based contracts), chronic disease management improvement (which affects risk scores and quality metrics), and patient satisfaction (which remote monitoring consistently improves for appropriate patient populations).

- Confirm that the RPM device vendor supports FHIR-compatible data export, or that an integration layer can map device output to standard FHIR Observation resources with LOINC codes.
- Define clinical alert thresholds collaboratively with the care team before deploying AI monitoring logic — alert parameters that are too sensitive generate

care team fatigue; parameters that are too conservative miss the clinical events the program is designed to catch.

- Establish clear on-call and escalation protocols for after-hours alerts; RPM data does not observe business hours, and an AI alert with no human response pathway is a patient safety gap, not a monitoring program.

- Evaluate equity impact before program launch: RPM programs that require patients to own smartphones, maintain broadband connectivity, and manage device charging introduce selection bias and may inadvertently exclude the highest-risk populations.

Diagram 7.4 – Remote Patient Monitoring: AI-Driven Alert Generation from FHIR Observation Data

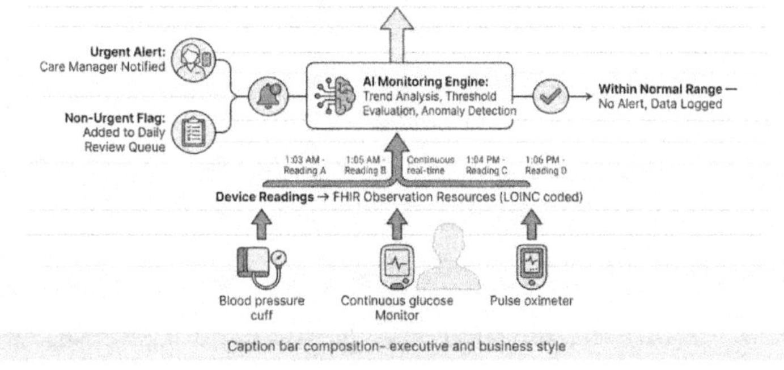

7.3.4 Patient Education

Patient education has long been recognized as a critical component of chronic disease management, medication adherence, and post-procedure recovery. The gap between what clinicians communicate during an encounter and what patients retain, understand, and act on is substantial. Studies consistently show that patients recall a fraction of what is discussed during appointments, and that generic written discharge instructions have minimal impact on behavior change.

AI-driven patient education, grounded in FHIR clinical data, addresses this gap by generating education content that is tailored to the specific patient's conditions, medications, care plan, and literacy level at the moment it is needed. When a patient is discharged with a new diagnosis of type 2 diabetes and a new prescription for metformin, an AI system can generate a personalized education package — what the diagnosis means, how the medication works, what side effects to watch for, what dietary changes will help, and when to call the care team — that is calibrated to that patient's health literacy level, language, and prior knowledge.

The FHIR resources that power this personalization include Condition (the diagnosis), MedicationRequest (the prescriptions), CarePlan (the care goals), and Patient (the demographic and language preference data). Large language models (LLMs) have proven effective at generating patient-facing explanations of clinical concepts in plain language, with appropriate reading level adjustment. When these models are given structured FHIR context rather than relying on general medical knowledge alone, the output is more accurate, more specific, and more clinically appropriate.

For managers, AI-generated patient education raises governance questions that must be answered before deployment. Who reviews the AI-generated content for accuracy? How are content updates managed when clinical guidelines change? How is the system prevented from generating content outside its validated scope — for example, answering clinical questions the patient asks that go beyond the education program's approved content? A robust patient education AI program requires content governance, clinician oversight protocols, and patient feedback mechanisms, not just a language model connected to a FHIR server.

7.4 Population and Public Health

Population health management and public health operations require a view of health that transcends the individual encounter. A clinician focuses on the patient in front of them. A population health program focuses on the entire attributed or enrolled population — identifying those most at risk, measuring care quality across all members, detecting emerging patterns before they become crises, and intervening at scale. Public health extends this further, encompassing community-level surveillance, outbreak detection, and policy-driven interventions.

FHIR is transforming population health by enabling the aggregation and analysis of clinical data across organizations, payers, and geographies without the fragmentation and delays of traditional data exchange mechanisms. When multiple health systems in a region share data through FHIR APIs — or when a health information exchange (HIE) operates as a FHIR-compliant aggregation layer — AI can work across the full population rather than the subset of patients any single organization can see. This multi-source, real-time data environment is qualitatively different from the claims-based, quarterly-refreshed datasets that most population health programs have historically relied on.

7.4.1 Predictive Modeling

Predictive modeling in population health uses historical clinical and administrative data to forecast future events: hospitalizations, emergency department visits, disease progression, care gap accumulation, and cost trajectory. These models support resource allocation, care management prioritization, network planning, and value-based contract performance management. The quality of a predictive model is fundamentally determined by the quality of the data it consumes.

FHIR-based predictive modeling has several distinct advantages over legacy approaches. First, the data is structured and standardized — FHIR resources use standard terminologies (SNOMED, LOINC, RxNorm, ICD-10) that make cross-organization and cross-system aggregation far more tractable than mapping disparate proprietary data structures. Second, the data can be accessed in near real time via FHIR APIs, allowing predictive models to update continuously rather than refreshing on a monthly claims cycle. Third, the FHIR data model is richer than claims data — it includes clinical observations, care plans, social history, and functional assessments that claims-based models cannot capture.

Health plans and ACOs using FHIR-enabled predictive models are reporting improvements in care management targeting accuracy — the right patients identified and enrolled in the right programs — compared with legacy claims-based approaches. The operational clarity this provides is significant: a care management program that knows it is reaching the top 5% of highest-risk patients with high precision can measure impact differently than one that relies on rough population cuts from quarterly claims data.

Managers overseeing predictive modeling programs should require transparency at the model level. Which features drive the prediction? What is the model's performance on the specific population being served, not just the training population? What is the prediction interval — how far in advance does the model generate useful signal? How does model performance degrade over time as the population's demographics and health status change? These questions define the model governance framework that responsible by design programs require.

7.4.2 Social Determinants of Health

Social determinants of health (SDOH) — housing, food security, transportation, economic stability, education, and social isolation — account for an estimated 30–55% of health outcomes. Clinical care, despite consuming the majority of health spending, addresses only a fraction of the factors that determine whether a patient stays well. Population health programs that ignore social determinants are working with an incomplete model of risk, and the interventions they design will be less effective than they could be.

FHIR has progressively improved its support for SDOH data. The FHIR R4 and R4B releases include Observation and Condition resources that can capture standardized SDOH screening results — food insecurity, housing instability, transportation barriers — using validated instruments like the PRAPARE, AHC HRSN, and Hunger Vital Sign screeners. The Gravity Project, a national collaborative that operates under the HL7 umbrella, has developed a comprehensive set of FHIR-compatible SDOH code sets that enable structured, comparable SDOH data across organizations.

AI systems that incorporate SDOH data alongside clinical data generate more accurate risk predictions and more mission-aligned interventions. A diabetes management AI that knows a patient has unreliable transportation is better positioned to recommend telehealth follow-up over in-person visits. A care manager AI that incorporates food insecurity data can flag patients for community resource referral rather than medication intensification when the likely driver of uncontrolled HbA1c is nutritional, not pharmacological.

The operational challenges of SDOH integration are real. SDOH screening rates are low in most clinical settings — patients who need resources most are sometimes least likely to complete screening tools. The community resource referral infrastructure is fragmented — electronic referrals to food banks, housing assistance programs, and transportation services require community health worker networks and closing-the-loop systems that many organizations have not built. For managers, building SDOH into population health

AI is a multi-year organizational development challenge, not just a data integration project.

Diagram 7.5 – Social Determinants Integration: FHIR SDOH Data Flow to AI Risk Model

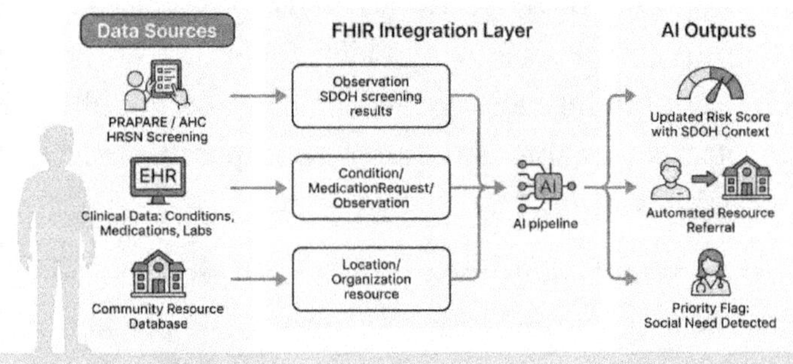

7.4.3 Community-Level Insights

Community-level insights represent the furthest extension of FHIR and AI's value — from the individual patient to the geographic community. Public health agencies, health systems with mission commitments to community benefit, and regional health information exchanges are the primary users of community-level analytics. The use cases range from disease surveillance and outbreak detection to community health needs assessment (CHNA) to health equity measurement.

FHIR-based disease surveillance is a capability that received significant attention during and after the COVID-19 pandemic. The limitations of traditional public health reporting — slow, manual, siloed by reporting entity — were dramatically exposed during the pandemic. FHIR-based reporting, where clinical systems automatically push case reports to public health agencies as FHIR resources in near real time, addresses these limitations directly. The HL7 Electronic Case Reporting (eCR) initiative has developed FHIR-based implementation guides for automated public health reporting that are now in production in multiple states.

AI applied to community-level FHIR data can identify disease clusters, detect unusual patterns in diagnosis or medication prescribing that may signal an emerging outbreak, and model the impact of population-level interventions. Geographic mapping tools that overlay FHIR-derived population health metrics with census data, environmental data, and community resource inventories provide public health planners with the analytical foundation for equity-focused resource allocation.

For health systems, community-level analytics support the CHNA process required for nonprofit hospital tax-exempt status under IRS 501(r) regulations. AI systems that can aggregate FHIR data across the health system's entire served population, segment it by geographic area and demographic group, and identify unmet health needs provide a more rigorous, more defensible CHNA than the manual survey and interviews processes that most organizations currently rely on.

The governance challenge at the community level is data sharing consent. Aggregating FHIR data across organizations for population health analysis requires patient consent frameworks, data use agreements, and de-identification standards that protect individual privacy while enabling population-level analysis. FHIR's built-in consent management resources — the Consent resource — and the emerging SMART Health Data Sharing framework provide technical mechanisms for managing this complexity. But the legal and governance frameworks must be established between participating organizations before the technical integration is attempted.

Diagram 7.6 – Community Population Health: FHIR-Enabled Multi-Organization Data Aggregation

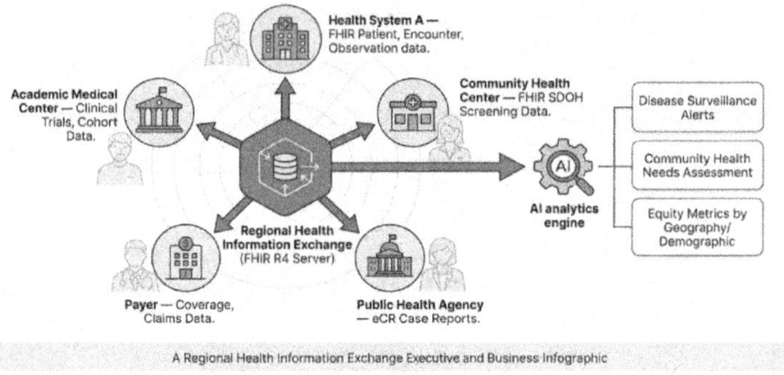

7.5 Takeaway

The use cases in this chapter are not aspirational. They are in production at health systems, payers, public health agencies, and technology vendors today. What distinguishes the organizations that are succeeding from those that are struggling is not the sophistication of their AI models — it is the quality of their data infrastructure and the rigor of their governance. FHIR provides the structured, interoperable, real-time data layer that AI needs to function reliably. Governance provides the accountability structures that ensure AI functions safely.

For managers, the chapter's practical value lies in sequence and selection. Not every organization is positioned to tackle all of these use cases simultaneously. The right starting point is the intersection of organizational readiness, clear ROI, and manageable risk. Administrative use cases — prior authorization automation, eligibility verification, revenue cycle optimization — typically offer the fastest path to measurable returns and the lowest clinical risk profile. Clinical use cases offer greater impact but require stronger governance infrastructure, clinical champion engagement, and validation rigor before go-live. Patient-facing and population health use cases often require the longest runway because they depend on data that is collected across multiple systems and sustained over time.

The pilot-to-production pathway for any use case in this chapter should follow a common discipline: establish baseline metrics before launch, define success criteria that are specific and measurable, include a bias and equity audit in the evaluation protocol, assign explicit human accountability for AI-generated outputs, and build a model monitoring function that will detect performance degradation over time. These are not optional steps — they are what separates responsible deployment from reckless adoption.

The organizations that will lead in the next decade of healthcare transformation are those building mission-aligned AI programs on FHIR foundations today. They are not waiting for perfect data, perfect models, or perfect regulations. They are making deliberate, sequenced investments in data infrastructure, governance, and workforce capability, and treating each production deployment as a learning system — something to be monitored, measured, improved, and held accountable. That discipline is more durable than any individual use case, and it is the foundation for sustainable, trustworthy AI in healthcare.

The next chapter examines the governance and compliance frameworks that underpin responsible AI deployment — the policies, processes, and organizational structures that convert operational clarity into institutional accountability.

8 Leading the Transformation: Governance, Trust, and the Human Side

Every chapter in this book has examined a technical layer of the FHIR-AI stack: interoperability standards, data pipelines, model deployment, security, and clinical integration. This final chapter turns to the layer that determines whether any of it actually works in practice: the human and organizational layer. The most technically sophisticated AI deployment in healthcare will stall, fail quietly, or cause harm if the people who operate it, govern it, and rely on it are not aligned, informed, and supported.

Leadership in healthcare AI is not primarily a technical role. It is an organizational and ethical one. The manager's job is to translate technical capability into mission-aligned outcomes, build the governance structures that keep intelligent systems accountable, and guide clinical and operational teams through the disruption that meaningful change always brings. This chapter provides the frameworks, decision tools, and leadership principles to do exactly that.

The tone of this chapter is intentionally direct. Healthcare organizations are entering a decade of deep AI integration. The decisions made now about governance structures, trust-building practices, and change management approaches will shape patient outcomes, staff experience, and institutional reputation for years to come. This is not a moment for aspirational language. It is a moment for clear-eyed operational planning and decisive leadership action.

8.1 The Human Barriers to AI Adoption

Technical readiness is rarely the primary obstacle to AI adoption in healthcare. Organizations that have invested in FHIR infrastructure, built clean data pipelines, and developed or procured capable models frequently discover that the harder problem is human: staff who distrust the technology, clinicians who resist workflow changes, administrators who cannot agree on accountability, and incentive structures that reward the old way of doing things. A manager who fails to diagnose and address these barriers will watch capable systems sit unused.

These barriers are not irrational. They are rooted in legitimate professional concerns, institutional history, and the genuine complexity of introducing machine judgment into high-stakes clinical environments. The leadership response must acknowledge that complexity honestly rather than dismissing resistance as ignorance or obstruction.

8.1.1 Fear of Job Displacement

The fear that AI will eliminate clinical and administrative roles is widespread and, in some cases, well-founded. Radiology, pathology, medical coding, prior authorization, and documentation have all been cited as areas where AI will reduce the number of humans required. Whether or not those predictions are accurate in any specific context, the fear itself is real and it shapes behavior. Clinicians who believe that an AI system is gathering evidence for their eventual replacement will not be enthusiastic partners in its deployment. They will find reasons for it to fail.

The manager's response is not to promise that no jobs will change — that would be dishonest and would destroy credibility when circumstances change. The response is to be specific about what the AI system is designed to do, what it is not designed to do, and what the organization's explicit commitment is to the people whose roles will be affected. Vague reassurances are worse than silence. Specific commitments — retraining pathways, role redesign timelines, explicit statements about what decisions remain human — are credible.

In practice, most healthcare AI systems deployed today augment rather than replace. An AI that flags abnormal imaging findings does not replace the radiologist; it changes the radiologist's workflow. The radiologist now reviews a pre-prioritized worklist with AI annotations rather than a raw, unsorted queue. That is a workflow change, not an elimination. But it requires the radiologist to develop new skills, adjust professional identity, and trust a system they did not design. Framing that honestly — as a workflow evolution with training support — is more effective than either overpromising or minimizing.

Specific actions managers can take to address displacement fear include: conducting role impact assessments before any deployment announcement, involving affected staff in workflow redesign rather than presenting finished solutions, creating visible career development pathways for AI-adjacent skills, and establishing a formal forum where staff can raise concerns without career risk. The goal is not consensus — some staff will remain opposed regardless — but informed participation. People who understand the plan and have had a chance to shape it are far more likely to support it than people who learn about changes on deployment day.

8.1.2 Lack of Clarity About Roles

When an AI system recommends a clinical action and the clinician follows that recommendation and something goes wrong, who is responsible? When a data science team builds a model using patient data and the model produces biased outputs, who is accountable? When an AI vendor's product is integrated into an EHR workflow and a patient is harmed, how is liability allocated between the vendor, the hospital, and the individual clinician? These questions do not have obvious answers in most healthcare organizations today, and that ambiguity is itself a barrier to adoption.

Clinicians are trained in accountability frameworks built around individual professional judgment. Introducing AI into that framework without a clear account of how responsibility is shared creates anxiety that manifests as resistance. If a physician believes that following an AI recommendation transfers liability to them — without corresponding protection or institutional backing — they will document dissent or simply override the system routinely, not because they believe the AI is wrong but because they are protecting themselves. That is a rational response to a governance failure.

Operational clarity on roles means defining, in writing and in policy, who owns each element of an AI-enabled workflow. This includes who is responsible for the model's training data quality, who validates model outputs before clinical use, who investigates adverse events involving AI, who decides when a model is retrained or decommissioned, and how individual clinicians are expected to interact with AI recommendations. These definitions do not eliminate uncertainty — AI governance is evolving rapidly — but they provide a working framework that staff can trust and build on.

The RACI framework — Responsible, Accountable, Consulted, Informed — is a useful starting point for mapping AI governance roles. Applied to a FHIR-based AI deployment, it might assign data engineering teams as Responsible for pipeline integrity, clinical informatics as Accountable for model validation, department chiefs as Consulted on workflow design, and frontline staff as Informed of model performance metrics. The specifics matter less than the clarity. When everyone knows their role, the human side of AI governance becomes manageable.

Diagram 8.1 – Human Barriers to AI Adoption and Leadership Responses

8.1.3 Resistance to Change

Resistance to change is not a personality flaw. It is a predictable organizational phenomenon with well-documented patterns. John Kotter's eight-step change model, William Bridges' transition framework, and the ADKAR model all describe the same basic reality: change requires a felt sense of urgency, a credible vision of the future state, and a supported pathway through the transition. Organizations that skip any of these steps encounter resistance that looks irrational from the outside but is entirely predictable from a change management perspective.

In healthcare AI, resistance typically concentrates at two points: initial announcement and the first incident. At announcement, staff who have seen technology projects fail, been excluded from planning, or had their workflows disrupted without benefit will assume the worst. At the first incident — a model flag that turns out to be wrong, a workflow slowdown caused by a new interface, an unexpected alert that interrupts a procedure — early skeptics will feel validated and neutral observers will move toward opposition.

The manager's job at both transition points is to demonstrate that the organization is responsive, not defensive. When resistance surfaces at announcement, the response is engagement: structured listening sessions, incorporation of feedback into design, visible evidence that staff input changed something. When resistance surfaces at an incident, the response is transparency: a clear account of what happened, what the investigation found, what changed as a result, and who is accountable. Organizations that handle the first incident well typically find that staff trust increases — not because nothing went wrong, but because leadership responded in a way that was honest and competent.

Pilot programs are the most effective tool for managing resistance at scale. A well-designed pilot in a single unit or service line creates a contained environment where problems can be surfaced and solved without organization-wide disruption. It also creates credible evidence — not vendor claims, not theoretical projections, but local data from the organization's own patients and staff — that the system works as described. The pilot-to-production pathway should be planned from the beginning, with explicit criteria for what constitutes a successful pilot and a defined escalation process for problems encountered during the pilot.

8.1.4 Misaligned Incentives

Healthcare organizations operate under complex, sometimes contradictory incentive structures. Volume-based revenue models may penalize efficiency gains from AI. Physician compensation structures that reward procedure counts may create resistance to AI tools that reduce procedure rates. Department-level budget accountability may make leaders reluctant to invest in organization-wide infrastructure that benefits others. These are structural issues, not attitudinal ones, and they require structural solutions.

The manager tasked with deploying an AI system to reduce readmission rates — and thereby reduce admission revenue in a fee-for-service environment — faces an incentive problem that no amount of change management will solve on its own. The solution requires executive alignment: a clear statement from organizational leadership that the incentive structure supports the AI initiative's goals, even when that creates short-term revenue tension. Without that alignment, frontline resistance to AI that hurts the department's numbers is not obstruction — it is rational self-protection.

Incentive alignment for AI adoption means identifying, for each major AI initiative, its primary beneficiaries and cost-bearers. In most cases, the cost of change — disrupted workflows, time spent learning new tools, short-term productivity loss — falls on frontline staff. At the same time, the benefits accrue to the organization as a whole or to a different part of the system. Closing that gap through recognition programs, adjusted productivity metrics during transition periods, and explicit acknowledgment of the burden frontline staff are carrying is not just fairness — it is a practical prerequisite for sustained adoption.

Where possible, create direct feedback loops between AI outcomes and the teams' performance metrics. If nursing staff are responsible for executing an AI-driven sepsis alert protocol, ensure that sepsis outcomes are visible in their unit metrics. If case managers are using AI to optimize discharge planning, ensure that length-of-stay and readmission data are part of their performance conversations. Mission-aligned incentives are not just financial — they are informational. Staff who can see the impact of their work stay engaged.

8.2 Building Trust in AI

Trust in AI is not a feeling. It is an organizational condition built through repeated, verifiable demonstrations that a system does what it is supposed to do, within the bounds defined, with appropriate human oversight in place. Trust is also fragile: it accumulates slowly through consistent performance and is lost quickly by a single visible failure poorly handled. The manager's role is to create the conditions under which trust can be earned and maintained, not to demand it as a precondition of deployment.

The healthcare context makes trust-building particularly consequential. Patients who learn that their care decisions were influenced by an AI system — particularly one they did not consent to, know about, or understand — may lose confidence in their care team and their institution. Clinicians who discover that a system they rely on has been producing biased outputs without their knowledge will lose confidence in both the technology and the organization that deployed it. The reputational and legal consequences of AI-related trust failures in healthcare are substantial and growing as regulatory scrutiny increases.

8.2.1 Transparency

Transparency in AI means making the system's behavior visible and understandable to the people who interact with it. It does not mean requiring every clinician to understand the mathematics of gradient descent or the architecture of a transformer model. It means that anyone affected by an AI system's output should be able to understand, at the level relevant to their role, how the system works, what it is optimizing for, its known limitations, and its current performance.

At the point of care, transparency means that AI-generated recommendations are labeled as such, that the confidence level or basis for the recommendation is available on request, and that the clinician has clear access to an override mechanism. The National Academy of Medicine and multiple professional societies have called for AI outputs in clinical settings to be clearly distinguished from human-generated content. This is a transparency minimum — the threshold below which trust cannot be established.

At the organizational level, transparency means regular, accessible reporting on model performance. This includes accuracy metrics, but also equity metrics — how does the model perform across patient demographics? What populations are underserved by its predictions? A model that performs well overall but performs poorly for elderly patients or patients from specific zip codes is not mission-aligned, regardless of its headline accuracy. Transparent reporting on equity disaggregation is increasingly required by regulators and expected by accreditors.

Explainability tools — SHAP values, LIME, attention visualization — allow technical teams to characterize why a model produced a specific output. These tools are valuable for internal validation but should not be confused with clinical transparency. A SHAP waterfall chart is not meaningful to a bedside nurse. Transparency at the clinical level requires translation: clear, plain-language explanations of what factors the model considered, presented in the context of the clinical workflow. Investing in that translation is a leadership decision, not an engineering task.

Diagram 8.2 – The Trust-Building Stack for Healthcare AI

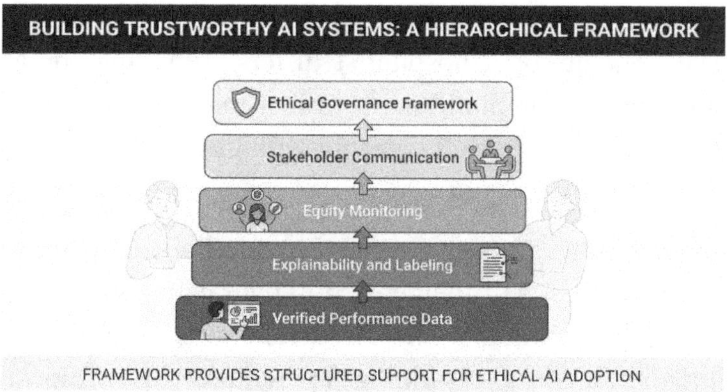

8.2.2 Communication

Communication about AI in healthcare organizations routinely fails in one of two directions: overselling or under-explaining. Overselling — framing AI as a transformative breakthrough that will solve systemic problems — creates expectations that operational systems cannot meet. When the technology performs within its actual capabilities but falls short of the hyped expectations, trust is damaged even though the system worked as designed. Under-explaining — deploying AI systems without communicating what they do, why they exist, and how staff should interact with them — creates suspicion and workarounds that undermine the system's effectiveness.

Effective AI communication follows a consistent structure. For each system being deployed, define and communicate: what the system does in plain language, what decisions it informs and what decisions remain human, what the system's known limitations are, how performance is being monitored, and who to contact with questions or concerns. This communication should reach every role that interacts with the system, tailored to that role's perspective. The content for a hospitalist differs from that for a care coordinator, which differs from that for an IT administrator, even when they are all working with the same system.

Patient communication is an area where most healthcare organizations are significantly behind. Patients have a right to know when AI is being used in their care, what it is being used for, and whether they have a choice about its use. Some jurisdictions are beginning to require AI disclosure in healthcare settings, and that trend will accelerate. Getting ahead of this requirement — establishing clear patient-facing communication about AI use before it is mandated — is both an ethical practice and a strategic one. Patients who understand and consent to AI use are more likely to trust the care system and less likely to generate complaints or litigation when something goes wrong.

Internal communication about AI performance should be regular, standardized, and honest. Monthly or quarterly AI performance reports, shared with clinical and administrative leadership, should cover model accuracy trends, alert fatigue rates, adverse events reviewed, and any corrective actions taken. These reports create accountability, build institutional knowledge, and provide documentation that demonstrates responsible stewardship to regulators and auditors. Organizations that can show a clean, consistent record of AI oversight will be far better positioned as regulatory scrutiny of healthcare AI increases.

8.2.3 Ethical Frameworks

Every healthcare organization deploying AI needs an explicit ethical framework that defines its values and commitments in this domain. This is not a philosophical exercise. It is an operational tool that guides decisions when the right course of action is not obvious — which is frequently the case in AI deployment. An ethical framework provides the reference point for questions like: Should we deploy a model that is accurate but unexplainable? Should we use patient data for model training without explicit consent if consent is impractical at scale? Should we accept an AI product from a vendor that cannot demonstrate performance equity across demographics?

The WHO, the AMA, the American College of Physicians, and major health systems, including Mayo Clinic, Cleveland Clinic, and Kaiser Permanente, have each published AI ethical frameworks. While the specific language varies, they converge on several core commitments: that AI must benefit patients, that AI must be fair across patient populations, that AI must be transparent and explainable to the extent possible, that human oversight must be maintained, and that privacy must be protected. These commitments are not aspirational — they translate directly into design requirements and governance policies.

An ethical framework for a healthcare organization should be developed with input from clinical ethics committees, legal and compliance teams, clinical staff, patient advocates, and technology leaders. It should be reviewed annually and updated as the regulatory and technical landscape evolves. Critically, it should be operationalized: each major AI deployment should include an explicit ethics review that checks the system against the organization's stated commitments. A framework that exists only as a document on the intranet is not a governance tool — it is a liability.

8.2.4 Responsible AI Principles

Responsible AI is not a certification or a product feature. It is a set of practices that an organization commits to and operationalizes across the full lifecycle of every AI system it develops or deploys. The phrase 'responsible by design' captures the most important insight: responsibility must be built into AI systems from the beginning of the development process, not audited in at the end. Organizations that attempt to retrofit responsibility onto fully developed systems consistently find it more expensive, more disruptive, and less effective than organizations that design for it from the start.

The core principles of responsible AI in healthcare are: safety (the system must not cause harm), efficacy (the system must do what it is claimed to do), equity (the system must perform consistently across patient populations), privacy (the system must protect patient data), transparency (the system must be understandable to relevant stakeholders), and accountability (there must be clear human ownership of the system's behavior). These principles map directly onto governance structures, which are covered in the next section.

One of the most important responsible AI practices is bias detection and mitigation. Healthcare datasets reflect historical inequities in care delivery — disparities in diagnosis rates, treatment access, documentation quality, and patient-provider interaction that vary systematically by race, ethnicity, gender, age, and socioeconomic status. Models trained on these datasets will encode those inequities unless explicit steps are taken to detect and mitigate them. This is not a theoretical concern. Documented cases of racial bias in clinical AI tools, including a widely cited algorithm that systematically underestimated the health needs of Black patients, demonstrate that bias in healthcare AI is a patient safety issue, not merely an ethics concern.

Responsible AI practices also require a clear process for model retirement. Systems that were accurate when deployed become outdated as clinical practice, patient populations, and data environments change. A responsible organization maintains a model inventory with performance monitoring and defined retirement criteria, rather than running models indefinitely because no one has taken ownership of the decision to decommission them. The decision to retire a model is a leadership decision, not a technical one: it requires weighing the cost of transition against the risk of continued use of an underperforming system.

8.3 Governance for Intelligent Interoperability

Governance is the organizational mechanism that keeps AI systems accountable over time. Without governance, even well-designed systems drift: models degrade as data distributions shift, security configurations erode as systems are patched and updated, compliance documentation falls out of date as regulations change, and no one takes responsibility for the widening gap between what the system was designed to do and what it is actually doing. Governance is not overhead — it is the mechanism that preserves the value of the investment.

For FHIR-based AI systems, governance operates at multiple levels simultaneously: the data level, the model level, the security and compliance level, and the operational monitoring level. Each level requires distinct policies, roles, and tooling. The manager's task is to design a governance structure that is comprehensive enough to catch real problems and lightweight enough to be sustained by people who have other responsibilities. Governance structures that require too much overhead are not followed; governance structures that require too little produce false confidence.

Diagram 8.3 – Governance Layers for Intelligent Interoperability

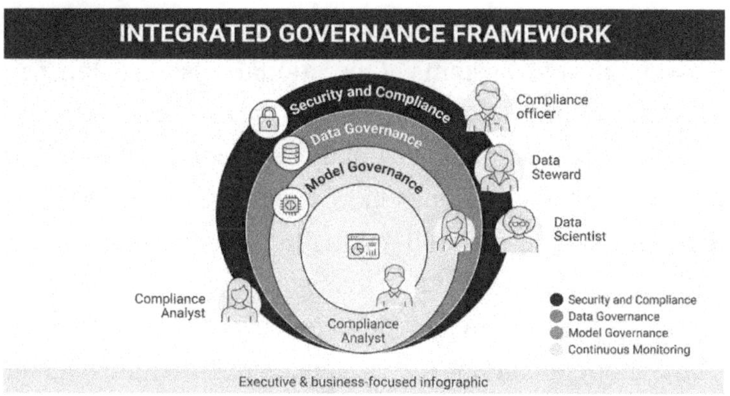

8.3.1 Data Governance

Data governance in the context of FHIR-based AI encompasses the policies, roles, and processes that ensure data quality, data lineage, consent management, and appropriate data use across the full pipeline from source system to model output. It is the foundation on which everything else depends: a model trained on poorly governed data is not a clinical tool — it is a liability. A pipeline that ingests data without documented consent is not an asset — it is a regulatory exposure.

The first data governance requirement is a comprehensive data catalog. This is a living inventory of every data source feeding AI systems, documenting: the system of origin, the data elements included, the FHIR resource types and profiles in use, the consent framework governing use of the data, the data steward responsible for quality, and the refresh frequency and mechanism. Organizations that cannot answer basic questions about where their AI data comes from are not equipped to govern their AI systems.

Data quality governance for AI is more demanding than for operational systems. Operational systems can tolerate occasional missing values or inconsistencies without significant consequence — a single incorrect phone number in a patient record does not affect care delivery. But AI models are sensitive to data quality in aggregate: systematic patterns of missing data, inconsistent coding, documentation variation across providers, and historical data entry habits all shape model behavior. Governance must include data quality monitoring specifically designed for AI use: completeness rates by FHIR resource type, consistency metrics across source systems, and alert thresholds that trigger investigation when quality falls below defined minimums.

Consent management is the most legally complex element of AI data governance. The use of patient data for AI model training, validation, and operation touches multiple regulatory frameworks: HIPAA's provisions for research and operations use, the evolving landscape of state AI transparency laws, FDA guidance on real-world evidence, and ONC rules on information blocking and patient data rights. The manager does not need to be a regulatory expert, but must ensure that legal and compliance counsel has reviewed the consent framework for each AI initiative and that consent documentation is maintained and auditable. The cost of getting consent management wrong — patient complaints, regulatory investigations, civil litigation — substantially exceeds the cost of doing it right.

Data governance also covers data lineage: the ability to trace any data element back to its source, through each transformation, to its eventual use in a model output. FHIR's structured, resource-based format makes lineage tracking tractable in ways that legacy HL7 v2 data often does not. Investing in lineage tooling early in the FHIR AI implementation is a governance best practice that pays dividends during audits, incident investigations, and model validation exercises. Organizations that can demonstrate complete data lineage are significantly better positioned with regulators and accreditors.

8.3.2 Model Governance

Model governance is the set of policies and processes that govern the lifecycle of AI models from development through validation, deployment, monitoring, and retirement. For healthcare organizations, model governance is increasingly a regulatory requirement rather than just a best practice. The FDA's framework for software as a medical device (SaMD) and the agency's AI/ML action plan require that high-risk AI models used in clinical decision support be governed with rigor equivalent to other medical devices. Even for AI systems that fall below the SaMD threshold, the governance practices required for regulatory compliance represent a sound operational baseline.

Model governance begins at development. Every model entering the organization's AI inventory — whether developed internally, procured from a vendor, or derived from an open-source foundation — should have a completed Model Card. This document, adapted from the framework developed at Google and widely adopted across the industry, specifies: the model's intended use, its training data characteristics, its performance metrics across patient subgroups, its known limitations, the validation process used, the responsible owner, and the conditions under which the model should not be used. The Model Card is the governance artifact that makes accountability specific.

Model validation for clinical use should be conducted in two phases. The first phase is technical validation: does the model perform as the developer claims, on data representative of the deployment population? This requires access to a validation dataset that is independent of the training data and drawn from the actual patient population where the model will be used. Models validated on academic medical center populations may perform quite differently in community hospital settings. The second phase is clinical validation: does the model's output, when used in the intended workflow, improve the clinical decisions it is designed to support? This is a higher bar, requires clinical involvement in the evaluation design, and is often not achieved before deployment because it requires prospective data. The gap between technical and clinical validation is a significant risk that governance must explicitly address.

Model versioning and change management are often overlooked elements of model governance. When a model is retrained — whether because of data drift, a vendor update, or a performance remediation — the retrained version should be treated as a new deployment, subject to the same validation and approval process as the original. Organizations that allow silent model updates — where the system in production changes without a formal governance review — are operating without effective model governance, regardless of what their policies say. Technical controls that flag model version changes and require documented approval before promotion to production are a minimum governance requirement.

8.3.3 Security and Compliance

The FHIR-AI stack introduces security risks that are qualitatively different from those of traditional healthcare information systems. FHIR APIs expose structured patient data to external applications and services in ways that legacy systems did not. AI models that ingest patient data at scale create aggregated datasets that are high-value targets for attackers. Model endpoints that receive clinical queries may be vulnerable to adversarial inputs designed to manipulate model outputs. Each of these attack surfaces requires specific security controls.

The SMART on FHIR authorization framework, covered in earlier chapters, is the primary mechanism for controlling FHIR API access. Governance requires that SMART on FHIR be implemented consistently across all AI-integrated FHIR endpoints, that OAuth scopes be defined with least-privilege principles, that token lifetimes and refresh policies be appropriate for clinical contexts, and that all API access be logged to an auditable record. These are not implementation details — they are governance requirements that must be specified in policy and verified through regular audit.

For AI systems that use Protected Health Information (PHI), the HIPAA Security Rule's requirements for access control, audit controls, integrity, and transmission security apply. But HIPAA's baseline requirements were designed for a pre-AI threat landscape. Organizations that rely solely on HIPAA compliance as their security framework for AI are underprotected. Supplementary controls should include: encryption of data at rest and in transit within AI pipelines, access controls on model training environments, automated detection of anomalous data access patterns, and regular penetration testing of AI-integrated interfaces.

Compliance for healthcare AI is a moving target. The FDA's evolving SaMD framework, the FTC's enforcement actions on AI claims, CMS guidance on AI in clinical decision support, the EU AI Act's extraterritorial provisions affecting US organizations with EU patient populations, and the growing body of state-level AI legislation all create a compliance landscape that is changing faster than most organizations can track. The governance response is to assign specific ownership of AI regulatory tracking — typically to legal and compliance leadership — and to build regulatory review into the annual governance calendar rather than treating compliance as a one-time deployment checklist.

Diagram 8.4 – AI Security and Compliance Architecture

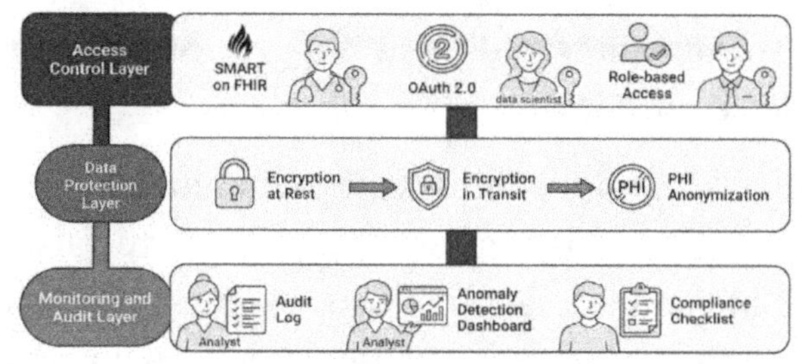

8.3.4 Continuous Monitoring

Deployment is not the end of governance — it is the beginning of the ongoing governance challenge. AI models in production require continuous monitoring for performance drift, data quality degradation, alert fatigue, clinical impact, and security anomalies. Organizations that treat deployment as the finish line and move on to the next project will find, months or years later, that the systems they deployed are running on stale models, producing outputs that bear little relationship to current clinical reality, and consuming organizational trust at a rate that is hard to recover from.

Performance drift is the most predictable monitoring challenge. Patient populations change, clinical practice patterns evolve, coding behaviors shift, and the data environment that a model was trained on gradually becomes less representative of the data it is receiving in production. This drift is not visible in individual predictions — any single model output may still look reasonable — but manifests in population-level metrics over time. Monitoring must include regular comparisons of current model performance against the baseline established at validation, with defined thresholds that trigger investigation and, if warranted, retraining.

Alert fatigue is a specific monitoring concern for clinical AI. Models that generate excessive false positives train clinicians to ignore alerts — a phenomenon called automation bias that is well-documented in aviation and increasingly studied in healthcare. Once alert fatigue sets in, it is difficult to reverse: the team has learned to dismiss the system and will continue doing so even when the false positive rate improves. Monitoring alert override rates is a leading indicator of alert fatigue, and governance should include a defined threshold — typically an override rate above 80-85 percent for any single alert type — that triggers workflow review before the system is further ignored.

Equity monitoring is a continuous requirement, not a one-time validation check. As patient populations evolve and data distributions shift, equity gaps that were not present at deployment may emerge. A model that performed equitably across demographic groups at launch may develop disparate performance as the makeup of the patient population changes or as documentation patterns vary across provider groups. Monthly equity monitoring by demographic group, with mandatory escalation when gaps exceed defined thresholds, is the governance standard that responsible organizations are moving toward.

Governance dashboards that provide consolidated visibility into all of these monitoring dimensions — performance metrics, alert rates, equity indicators, security alerts, compliance status — are an investment that pays for itself in management efficiency and regulatory readiness. These dashboards should be accessible to clinical and operational leadership, not just data science teams. When governance visibility is confined to technical staff, organizational leadership cannot fulfill their accountability obligations. The goal is operational clarity at every level of the organization that has accountability for AI performance.

8.4 Leading Through Change

Healthcare AI is not a project with a defined end state. It is a long-running organizational transformation that will produce a succession of new capabilities, new governance challenges, and new human adaptation requirements for the foreseeable future. Leaders who approach it as a project — with a launch date, a go-live, and a close-out report — will be repeatedly surprised when the complexity does not diminish after deployment. Leaders who approach it as a transformation — ongoing, iterative, requiring sustained organizational investment in people and culture — will build the institutional capacity to navigate what comes next.

The leadership skills most relevant to healthcare AI transformation are not primarily technical. They are the skills of organizational navigation: building coalitions across organizational boundaries, communicating complex uncertainty in honest ways without being paralyzing, holding people accountable while preserving the psychological safety that learning requires, and making resource allocation decisions under conditions of genuine uncertainty. These are the capabilities that will determine which organizations successfully integrate AI into their care delivery model and which ones accumulate failed deployments and eroded trust.

8.4.1 Guiding Teams Through Uncertainty

Healthcare professionals are trained to resolve uncertainty as quickly as possible. Clinical decision-making frameworks, diagnostic algorithms, and evidence-based guidelines all exist to reduce the space of uncertainty within which practitioners must operate. AI introduces a different kind of uncertainty — probabilistic, calibrated, and explicitly incomplete — that sits uncomfortably with clinical training. A model that says 'this patient has a 73 percent probability of sepsis within the next six hours' is not a diagnosis. It is a structured uncertainty estimate. Learning to act appropriately on that kind of output requires a cognitive shift that takes time and deliberate practice.

The manager's role in guiding teams through this uncertainty is to name it clearly and provide structured tools for working with it. This means explicit training on how to interpret probabilistic outputs, case discussions that walk through how AI recommendations should influence — but not replace — clinical judgment, and a clear organizational position on the threshold of AI confidence required before a recommendation should be acted on. Organizations that leave clinicians to figure out how to use probabilistic AI outputs on their own will get inconsistent results and, eventually, an adverse event attributable to misunderstood AI guidance.

Teams also need guidance for situations where the AI system's output conflicts with clinical intuition. This happens, and it will continue to happen. The governance question is: what is the right default? Should the clinical override require documentation? Should it trigger a secondary review? Should it feed back into model improvement processes? Organizations that design explicit protocols for AI-clinician disagreement create both a safety mechanism and a valuable data stream: the cases where experienced clinicians override AI recommendations are often the cases where the model has encountered a population or presentation it is not well-calibrated for.

Uncertainty is also institutional. The regulatory framework for healthcare AI is still developing. Reimbursement models for AI-enabled care are inconsistent and evolving. Evidence for the real-world clinical impact of specific AI tools is accumulating but incomplete. A leader who presents AI transformation as a certain path to defined outcomes is misleading their team. The honest framing is: 'We have strong evidence that AI can improve specific outcomes in specific contexts. We are building the governance and implementation capability to capture that value responsibly, and we will adjust our approach as we learn.' That kind of honest, adaptive leadership is what sustains organizational commitment through the inevitable setbacks of transformation.

8.4.2 Creating a Culture of Innovation

A culture of innovation in healthcare AI is not a culture of uncritical enthusiasm for new technology. It is a culture that combines intellectual curiosity with rigorous skepticism, rapid experimentation with disciplined evaluation, and tolerance for failure with accountability for learning. This combination is difficult to build and easy to destroy. Leaders who punish failure create cultures that never experiment. Leaders who celebrate innovation without enforcing rigor create cultures that generate impressive pilots that never make it to production.

The structural enabler of innovation culture is the pilot-to-production pathway. When teams know that a successful pilot has a defined and resourced path to broad deployment, they are motivated to invest seriously in the pilot. When they have seen previous pilots succeed technically but die in the transition to production — because of budget, political resistance, or organizational priority changes — they stop believing in the process and stop innovating. The most important cultural investment a healthcare AI leader can make is building a credible, transparent, and consistently followed process for moving innovations from idea to deployed system.

Innovation culture also requires dedicated space for experimentation. This does not mean unlimited resources — healthcare organizations cannot afford open-ended exploration without defined goals. It means protected capacity: time for clinical informatics staff to explore new AI applications, a sandbox environment where models can be tested against de-identified data without production consequences, and a clear process for proposing new AI initiatives that is accessible to clinical staff, not just technology teams. Some of the most valuable healthcare AI insights come from clinicians who understand their workflow in ways that no data scientist can, if they are given a structured way to contribute to AI development.

Measuring innovation culture means tracking the health of the pipeline, not just the outcomes of deployed systems. How many new AI initiatives are proposed each year? How many advance from proposal to pilot? How many advance from pilot to production? What is the average time from pilot success to production deployment? If the pipeline is thin at any stage, the culture has a problem at that stage. Leaders who only measure post-deployment outcomes are looking at a lagging indicator that will not tell them where the innovation process is breaking down until significant opportunity has been lost.

Diagram 8.5 – Innovation Pipeline: Idea to Production

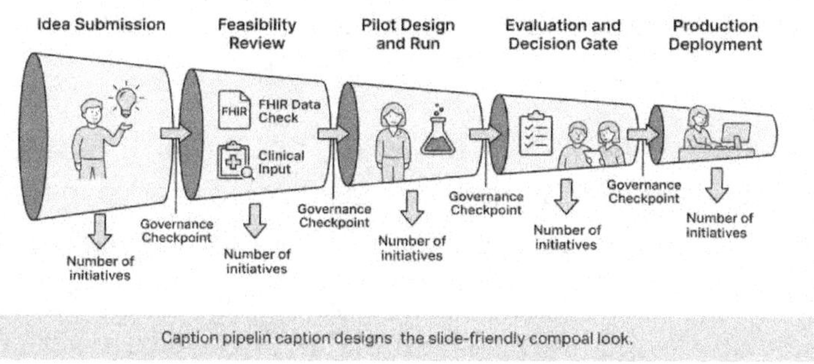

8.4.3 Empowering Clinicians and Staff

The most durable form of AI adoption is adoption that clinicians and staff own. When clinical teams are involved in problem identification, workflow design, validation, and ongoing monitoring, they become advocates rather than subjects of AI deployment. This is not merely a change management strategy — it is a quality imperative. Clinicians bring domain knowledge that no data scientist or technology leader can replicate. AI systems designed without deep clinical input consistently miss workflow realities that determine whether the system is actually used.

Clinical AI champions — clinicians who are trained in AI fundamentals, engaged in governance processes, and trusted by their peers — are among the highest-leverage investments a healthcare AI leader can make. A clinical AI champion is not an IT liaison or a super-user in the traditional sense. They are a peer leader who can translate between clinical and technical languages, identify workflow-level impact before deployment, troubleshoot adoption problems in real time, and give credible feedback on model performance from the perspective of someone who uses the system daily. Champion programs that invest in formal training, protected time, and recognition produce returns that are disproportionate to their cost.

Staff empowerment also means giving frontline teams the data and tools to understand AI performance in their own context. A sepsis prediction model that performs at 85 percent sensitivity for the hospital overall may perform at 72 percent for the night shift in the medical ICU, where patient acuity patterns, documentation timing, and monitoring frequency are different. If the MICU night team has access to that unit-specific performance data, they can have a meaningful conversation about what the model is and is not telling them and how to integrate it appropriately into their practice. If they only have access to hospital-wide statistics, they are being asked to trust a system without the information needed to calibrate that trust.

The empowerment framework for healthcare AI has three components: knowledge (staff understand what the system does and how to interpret its outputs), agency (staff have meaningful input into how the system is used in their context and a clear mechanism for raising concerns), and accountability (staff are expected to use the system appropriately and are supported in doing so, but also held to account when they systematically ignore or misuse it). This framework treats clinical and administrative staff as professional partners in AI governance rather than end-users of a technical product, which is both more respectful and more effective.

8.4.4 Preparing for the Next Decade of AI-Enabled Healthcare

The AI capabilities available to healthcare organizations today are a fraction of what will be available in five to ten years. Large language models are already demonstrating clinical reasoning capability that, while not yet deployable without supervision, is advancing at a rate that will make current limitations obsolete within this decade. Multimodal AI that integrates structured EHR data, imaging, genomics, and clinical notes into unified patient representations is moving from research to early clinical deployment. Ambient intelligence — AI that observes clinical encounters and generates documentation, follow-up recommendations, and care coordination actions without requiring explicit clinician input — is already in limited production use and expanding rapidly.

The organizations best positioned to deploy these capabilities effectively are not necessarily those with the most advanced AI today. They are the ones building the governance structures, data infrastructure, clinical trust, and leadership capabilities now that will allow them to adopt responsibly when the next generation of technology arrives. The FHIR infrastructure investments described throughout this book are foundational to that readiness: organizations that have clean, standardized, interoperable data pipelines will be able to integrate new AI capabilities far faster than those still managing fragmented legacy data environments.

The workforce implications of AI over the next decade are significant and require proactive planning now. The skills that will be most valuable in AI-enabled healthcare are not primarily clinical informatics skills — those are a means to an end. The most valuable skills will be the ability to critically evaluate AI systems, govern them responsibly, design workflows that integrate human and machine judgment appropriately, and lead organizations through sustained technological transformation while maintaining the quality, safety, and humanity of care. These skills are not taught in most clinical or administrative training programs today. Building them requires intentional investment.

Healthcare leaders should be planning now for a clinical environment in which AI is as embedded in care delivery as imaging or laboratory testing. That does not mean accepting AI uncritically or moving faster than safety allows. It means building institutional capacity — governance, data infrastructure, workforce capability, and leadership commitment — to integrate AI at the scale and sophistication required for patient care. The organizations that treat this as a one-time technology project will not be ready. The organizations that treat it as an ongoing institutional transformation will be.

Diagram 8.6 – The AI-Ready Healthcare Organization: Capability Maturity Model

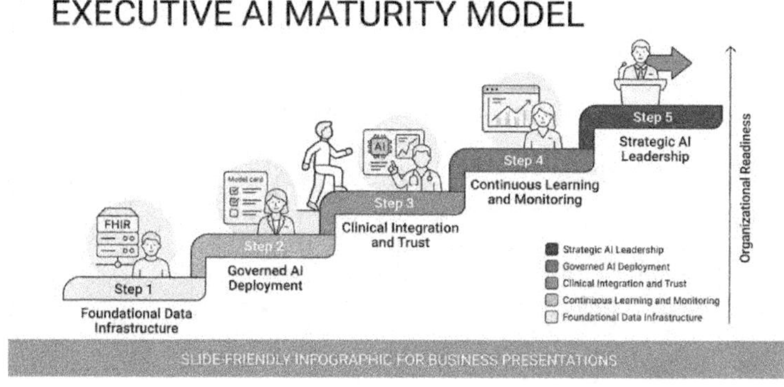

The maturity model above is not a prescription — no organization will move through these stages in a perfectly linear fashion, and the specific capabilities at each stage will vary by organizational context. It is a tool for strategic self-assessment: where is the organization today, what is needed to advance to the next stage, and how should investment be prioritized? Leaders who can answer those questions with evidence — not aspiration — are the ones who will guide their organizations through the next decade of AI-enabled healthcare with their patients' trust and their institutions' reputations intact.

The relationship between FHIR and AI is ultimately a relationship about information flow — how data generated in the process of caring for patients can be structured, governed, and applied to improve the quality and consistency of that care. FHIR provides the interoperability foundation. AI provides the analytical capability. But the transformation only delivers value when the human and organizational layer — the governance, the trust, the leadership, and the culture — is in place to deploy it responsibly. That is the work this chapter has described. That is the work that begins, not ends, with deployment.

8.5 Takeaway

Chapter 8 has covered the organizational dimensions of healthcare AI transformation: the human barriers that must be addressed, the trust-building practices that create durable adoption, the governance structures that maintain accountability, and the leadership approaches that sustain change over time. These are not soft topics — they are the hard work of making technically capable systems actually deliver value in complex, high-stakes clinical environments.

The Human Barriers Are Predictable and Addressable

Fear of job displacement, role ambiguity, resistance to change, and misaligned incentives are the four human barriers that most reliably stall healthcare AI adoption. None of them is insurmountable. Each requires a specific leadership response: honest communication about role impact, written policy clarity on accountability, structured change management with a credible pilot-to-production pathway, and incentive alignment that connects individual performance to AI-enabled outcomes. The organizations that address these barriers proactively outperform those that discover them after a failed deployment.

Trust Is Earned Through Consistent, Transparent Performance

Building trust in AI requires operational practices, not assurances. Transparency at the point of care, equity monitoring across patient demographics, structured communication plans for every major deployment, and an ethical framework that is operationalized rather than aspirational are the practices that earn trust from clinicians, patients, and regulators. Responsible by design means building these practices in from the start of every AI initiative, not auditing for them after deployment.

Governance Operates at Four Levels

Effective governance for intelligent interoperability requires policies, roles, and monitoring at the data level, the model level, the security and compliance level, and the operational monitoring level. Data governance ensures that the pipeline feeding AI systems is clean, consented, and documented. Model governance ensures that systems are validated, versioned, and monitored for drift. Security and compliance governance ensures that the FHIR-AI stack meets the evolving regulatory requirements of healthcare. Continuous monitoring ensures that all of these elements remain effective after deployment.

Leadership Through Change Requires Specific Skills

Guiding teams through uncertainty, building an innovation culture that produces a healthy pilot-to-production pipeline, empowering clinical AI champions, and preparing the workforce for the next decade of AI-enabled care are leadership tasks, not technology tasks. The managers who will successfully lead healthcare AI transformation are those who invest as much in organizational capability — governance structures, workforce development, cultural norms — as they do in technical infrastructure.

The Work Starts Now

The capabilities that will define AI-enabled healthcare in the next decade are being built on the foundations that organizations put in place today. FHIR infrastructure, model governance policies, clinical trust programs, and leadership capability are all long-lead investments: they take years to build and cannot be acquired quickly when they are suddenly needed. The organizations that will lead in AI-enabled healthcare are not waiting for technology certainty before they build these foundations. They are building now, learning as they go, and developing the institutional resilience to navigate whatever the next generation of AI capability brings.

1. Conduct a human barrier assessment for your most critical AI initiative — map displacement fears, role ambiguities, incentive conflicts, and resistance sources before the deployment announcement.
2. Develop a written AI ethical framework with input from clinical ethics, legal, compliance, and patient advocacy, and schedule an annual review.

3. Build a model inventory with completed Model Cards for every AI system currently in production or under evaluation, including vendor-supplied tools.
4. Implement equity monitoring dashboards that disaggregate AI model performance by race, ethnicity, age, gender, and socioeconomic proxies, with defined escalation thresholds.
5. Establish a clinical AI champion program with formal training, protected time, and a defined role in governance and model evaluation.
6. Define and document your pilot-to-production pathway: the criteria for a successful pilot, the resource commitment for production deployment, and the escalation process for problems encountered at each stage.
7. Assign ownership of AI regulatory tracking to a named compliance leader and build regulatory review into the annual governance calendar.
8. Assess your organization's position on the AI capability maturity model and identify the specific investments needed to advance to the next stage over the next 12 to 24 months.

The FHIR-AI relationship delivers value only when the technical foundation and the human organization are both fit for purpose. The technology is ready. The governance frameworks are available. The clinical demand is real. What remains is leadership — clear-eyed, mission-aligned, and prepared for the long work of making AI earn its place in the systems that care for people.

9 Conclusion

Healthcare stands at a crossroads where the decisions leaders make today will determine whether AI becomes a force for safety, equity, and operational excellence—or a source of new risk, fragmentation, and mistrust. Throughout this book, one message has been constant: AI does not succeed in healthcare by accident. It succeeds when leaders intentionally build the data, governance, and human-centered structures required for intelligence to operate safely at scale. As the introduction reminds us, *"Most organizations didn't decide to become 'AI organizations.' It crept up on them."* The conclusion is different: becoming a **responsible AI organization** must now be a deliberate choice.

FHIR is the backbone of that choice. It is the standard that transforms healthcare data from a liability into an asset, from a scattered collection of incompatible systems into a unified, machine-readable ecosystem. AI, for all its promise, cannot overcome poor data quality, inconsistent coding, or inaccessible clinical records. But when paired with FHIR, AI becomes something fundamentally more powerful: a system that learns continuously, adapts safely, and operates with transparency. This book has shown that the future of healthcare is not AI alone, nor FHIR alone, but the **symbiosis** between them.

As you move forward, the most important shift is recognizing that interoperability is no longer a compliance requirement—it is a strategic differentiator. Organizations that treat FHIR as a checkbox will struggle with the same fragmentation, administrative burden, and model failures that have plagued the industry for decades. Organizations that treat FHIR as an intelligence platform will unlock capabilities that were previously impossible: real-time clinical insights, automated workflows, predictive care pathways, and patient-facing tools that feel intuitive rather than burdensome. The leaders who understand this distinction will define the next decade of healthcare innovation.

But technology alone is not enough. AI requires trust, and trust requires governance. Human-in-the-Loop (HITL) oversight is not a barrier to innovation—it is the mechanism that ensures AI remains aligned with clinical judgment, ethical standards, and organizational values. As the introduction states, HITL ensures humans *"actively review, validate, override, and improve"* AI decisions. This is not optional. It is the foundation of safe deployment. The organizations that thrive will be those that design oversight frameworks that scale, that calibrate human involvement intelligently, and that treat governance as a living system rather than a static policy.

The human side of this transformation is equally critical. AI will not replace clinicians, but it will absolutely reshape their work. Leaders must prepare their teams for that shift—not with fear, but with clarity. Clinicians need to understand how AI works, when to trust it, when to question it, and how it fits into the broader mission of patient care. They need tools that reduce burden, not add to it. They need workflows that feel intuitive rather than imposed. And they need leaders who communicate openly about what AI can and cannot do.

The same is true for patients. AI-enabled care must enhance transparency, not obscure it. It must improve access, not widen disparities. It must strengthen relationships, not weaken them. FHIR plays a central role here as well: it gives patients access to their own data, enables app ecosystems that support self-management, and ensures that intelligence follows the patient across settings rather than remaining locked inside institutional walls.

As you close this book, the path forward becomes clear. The organizations that will lead the next era of healthcare are those that:

- **Invest in FHIR not as an IT project, but as a strategic platform.**

- **Build AI systems that are explainable, governed, and continuously monitored.**

- **Design HITL workflows that protect safety without slowing innovation.**

- **Treat data quality as a clinical and operational priority.**

- **Empower clinicians and patients with tools that enhance—not replace—human judgment.**

- **Adopt interoperability as a cultural value, not a regulatory burden.**

The transformation ahead is not simple, but it is achievable. You now have the frameworks, the vocabulary, and the strategic lens to lead it. The question is no longer whether AI will reshape healthcare—it already has. The question is whether your organization will shape that future intentionally or be shaped by it.

If you choose the former, FHIR will be your foundation, AI will be your accelerator, and governance will be your compass. Together, they form the architecture of intelligent interoperability—the architecture of a healthcare system that learns, adapts, and improves every day.

Your role as a manager, leader, or strategist is not to predict the future. It is to build the conditions under which the future can be trusted. With the insights from this book, you are now equipped to do exactly that